GEOLOGY AND MINERALOGY RESI

ADVANCES IN GEOCHEMISTRY RESEARCH

Geology and Mineralogy Research Developments

Additional books in this series can be found on Nova's website under the Series tab.

Additional e-books in this series can be found on Nova's website under the e-book tab.

Earth Sciences in the 21st Century

Additional books in this series can be found on Nova's website under the Series tab.

Additional e-books in this series can be found on Nova's website under the e-book tab.

GEOLOGY AND MINERALOGY RESEARCH DEVELOPMENTS

ADVANCES IN GEOCHEMISTRY RESEARCH

JORGE SANJURJO-SÁNCHEZ
EDITOR

NOVINKA

New York

For permission to use material from this book please contact us:
Telephone 631-231-7269; Fax 631-231-8175
Web Site: http://www.novapublishers.com

Library of Congress Cataloging-in-Publication Data

ISBN: 978-1-62618-245-5

Published by Nova Science Publishers, Inc. † New York

Contents

PREFACE

Jorge Sanjurjo-Sánchez
University Institute of Geology, —Isirb Parga Pondal,”
University of A. Coruña, Edificio Servixos Centrais de Investigación,
A. Coruña, Spain

Geochemistry studies the chemistry of the Earth. It can be considered as a discipline that deals with the distribution and migration of the chemical elements within the Earth. Thus, it is different that other disciplines usually considered as similar, such as petrology or mineralogy. However, geochemistry comprises much more defined challenges, such as the study of the origin of chemical elements, their evolution, the formation of minerals and rocks minerals, and even the role of elements in biology (geobiology). We can even do some divisions in this discipline such as consider organic and inorganic geochemistry. Therefore, geochemistry today includes the study of both the interior and crust of the Earth, the composition of the atmosphere, interaction with living organisms, and even the distribution of elements in the Solar System and other planets. Moreover, other disciplines have taken the knowledge developed in geochemistry for different purposes. As an example, archaeology has used the geochemistry of elements analysed in ancient objects to assess the origin of raw and manufactured materials in the past.

Form the beginnings the geochemistry has undergone very important advances, above all in the twentieth century. Most advances have been related with the development of advanced instrumental techniques, geochemical modelling and geochemical exploration. New, instrumental techniques have improved the results of geochemical analysis and improved detection limits and the reliability of multi-element analytical methods. Some of the developed instrumental techniques are today routinely applied. Improvements in sampling techniques, development of 3-D models to assess results, use of new statistical techniques and numerical modelling aimed the refinement of geochemical models and knowledge on processes.

Modern geochemistry can be applied to nearly all of the Earth Sciences. In my opinion it is impossible to include all the possible applications of geochemistry in a single book. Therefore this book is an attempt to show and spread some of the last advances in the use of geochemistry in different subjects. The book includes chapters dealing with the use of geochemistry in environmental geochemistry, correlation of geological materials, use of rare earths, study of clays, biogeochemistry of petrological applications of geochemistry.

Chapter 1 includes the first geochemical analysis of the formations that conform the Pleistocene-Holocene geological record of the Doñana National Park, the most important Biosphere Reserve in Spain. In the last decade, numerous studies were focused on the geochemical analysis of estuaries with potential pollution sources. Such studies are especially important if a determined estuary contains a high biodiversity, as it is interesting to obtain a regional geochemical background in order to clarify the impact of several human-induced activities on soils or sediments. This chapter analyzes the geochemistry of recently described sandy formations of the Park to obtain an additional geochemical background for these coarser sediments. The authors propose a multiple geochemical background, with inclusion of unpolluted samples derived from a sediment core collected in the northwestern sector of this area. This background was used to evaluate the impact of the 1998 Aznalcóllar mining spill.

In Chapter 2 a comparative study of mineralogical and geochemical data of Pliocene clay deposits collected in west Portugal is presented. Quartz is usually the most common mineral in the deposits and occurs in larger amounts in the coarser sediments north of Mondego River. Clays such as Illite are always higher in the basins to the north of the Mondego, while kaolinite tends to be higher in the southern basins, where it is often the most common mineral. One set of samples from lower stratigraphic position of the northern sector (Vale Grande) has mineralogical features that resemble the southern deposits. In accordance with the mineralogical composition, silica and alumina are the most common oxides. Silica tends to be higher in coarser samples, in particular in the deposits south of Mondego River. Potassium oxide content is always higher in the northern than in the southern deposits. The observed differentiation between clay deposits is attributed to provenance. The deposits located to the north of Mondego River were derived mainly from the Precambrian metapelites of the Central Iberian Zone. Regarding the deposits to the south of Mondego River, the dominance of kaolinite and high Al_2O_3 content suggest that they result mainly from recycling of Cretaceous and Cenozoic units, which, in turn, derived from weathering and erosion of granitoids and metasedimentary rocks. Some older deposits north of Mondego were also originally sourced by previous cycle sediments, but provenance must have shifted to basin-edge metapelitic rocks due to uplifting of coastal mountains, which blocked the drainage from the hinterland. Besides provenance, hydraulic selection could explain local compositional differences.

In Chapter 3 it is presented a review on the geochemical evolution of materials surfaces in the built environment, focusing namely on the geochemical diversity of decay products. The same geochemical weathering processes that affect natural rock outcrops act on stony materials after their emplacement on the built structures (enhanced by higher concentration of pollutants related to the urban environment and to specific geometric configurations related to architectural works) leading to the formation of alteration products that could involve the mobilization of substances of the materials and the formation of new substances (that were not present on the materials before emplacement). Two major categories of alteration products will be considered in this review: coatings and erosive features. Coatings can result from neoformations related to pollutants and the reactions between pollutants and the substrate. They could also involve the fixation of exogenous matter by the surfaces and pores of the materials and by the neoformations substances. In erosive features will be included processes that imply the loss of material or that create conditions for that loss (fissuring). They can result from several processes that involve the chemical instability of the original phases of the materials, as well as physical consequences related to the interaction with pollutants, such as expansion and crystallization inside the pore network. The alteration products of stony materials in the built environment present specific geochemical aspects related to the specific conditions and its study is paramount for the definition of qualitative and quantitative relationships between pollutants and alteration features and this review aims to a synthesis in relation to these features.

Chapter 4 deals with the formation of granites in the Archaean – Proterozoic Boundary. The Archaean crustal provinces of the World are dominated by the rocks of the Tonalite-Trondhjemite-Granodiorite (TTG) series. Towards the end of Archaean, however, TTG magmatism was replaced by the generation of high-K granites which continued into the Proterozoic and Phanerozoic. The late Archaean high-K granites of the Tanzania Craton some of which intrude the greenstone sequence of northern Tanzania are compared with the early Proterozoic high-K granites from the Ubendian belt of southwestern Tanzania. Granites from both late Archaean and early Proterozoic crustal terranes are pink in colour, and are composed of K-feldspar, plagioclase, quartz, biotite, less hornblende with minor phases including sphene, zircon and chlorite. It is in the Proterozoic granites that muscovite and tourmaline become important phases.

The compositional similarities of the high-K granites from both late Archaean and early Proterozoic terranes include high SiO_2, K_2O and low MgO and Ni. Trace element compositions reveal that the high-K granites from both crustal terranes show comparable concentrations of Y, MREE and HREE, and display the negative Eu anomaly. The major compositional differences between them are discerned in the most incompatible elements. High-K granites from the Proterozoic terrane consistently show higher concentrations of Zr, Hf, Th, Ba, Sr, K and $\sum$REE than their late Archaean counterparts. On the basis of the observed compositional similarities and differences, the key trace element ratios employed to distinguish the two magmas involve an incompatible element and a middle to heavy REE (i.e. Zr/Sm, Zr/Y, Zr/Nb, Ba/Nb, Th/Ta) as well as the K_2O/Na_2O ratio. These ratios are a reflection of the primary petrogenetic feature inherent from the source rocks.

The field relationships and geochemical features of the high-K granites from both crustal terranes are consistent with their derivation from partial melting at pressures and depths where plagioclase was a residual phase. Whereas the generation of the high-K granites in the Archaean would require partial melting of the juvenile igneous rocks including the TTG and greenstones that were emplaced in previous 10 to 20 Ma, those in the Proterozoic would require partial melting of the more evolved crustal rocks including the Archaean rocks. The generation of high-K granitic magmas at the Archaean-Proterozoic boundary marked a transition from growth of the Neoarchaean continental crust through the addition of juvenile mantle-derived material to intra-crustal recycling of pre-existing material. It also marked the final stages of the stabilization of the Archaean Cratons worldwide.

Chapter 5 is an overview of the microbial mineralization in caves. Caves are windows into the subsurface, which allow scientists to explore the space below the Earth's surface. Most common types of caves are those formed in limestone and other calcareous rocks, as well as lava tubes in basaltic rocks. Remaining types include those formed in gypsum, quartzite, sandstone and granite, which are usually limited in extent and, in some cases, rarely found and investigated. Biomineralization, the processes by which organisms form minerals (from bioorganic molecules and inorganic solids), is receiving renewed attention by geochemistry and other disciplines. Microorganisms inhabit all possible biosphere ecosystems including subsurface rock environments, such as caves, mines, tunnels, etc., Such environments are nutrient-limited and contain a wide variety of redox interfaces and microorganisms able to promote biomineralization processes. By interacting with minerals, microorganisms play an important role in reshaping the mineral environment, and may help to form features such as mineral deposits and some types of speleothems in cave environments.

Chapter 6 shows and describes some important advances of applied geochemistry in China. The Regional Geochemistry National Reconnaissance (RGNR) program initiated in 1979 has covered about 7.7 million (80%) of China's territory by within-cell stream sediment sampling. The results will soon be published in an Atlas for the distribution of 39 elements in China both in English & Chinese. The tremendous amount of 39 element data has led to establish the geochemical block concept for predicting world-class mineral deposits, and to the discovery of about 1900 economic mineral deposits. A program of Geochemical mapping of 76 elements (96GEM) using the composited samples of South & SW China derived from the RGNR sample bank was finished and will soon be published as an atlas both in English & Chinese. The multi-purpose regional geochemical soil survey program initiated in 1999 had covered 1.65 million km^2 in Eastern China. The Geochemical Atlas of Hainan Island, Panyang Basin of Jiangxi and Pearl River delta were published for environmental, agricultural and ecological studies. A program of studying nearly all element contents in the periodic table in all the ore tailing dumps in China was initiated.

In order to overcome the general shortcomings of weak selective extraction used in Deep penetrating geochemistry new approaches have been tested including: ligand based extraction, strong acid extraction of micro-particles separated from samples, and strong acid selective extraction of the stable end products of variable oxides and sulfides after the mobile unstable form of element are moved to the surface through exotic covers.

Large orientations studies for surface geochemical and microbiological surveys of oil and gas were initiated in Heilongjiang province including in region where the famous Daqing oil field located and also in Sichuan Province and Tibet. Geochemical surveys for oil and gas began to be used in China in the 1980's, and nearly stopped at the end of 20th century. A new large program of geochemical exploration for oil and gas to cover whole China was worked out. It could be started after the success of the orientations studies. Chinese geochemists had been active in the international research program IGCP 259 for standardization of the multi-element analysis system, and IGCP 360 in the study of sampling methodology for the global-scale geochemical mapping. In view of the slow progress of global scale geochemical mapping, new concepts based on the fractal self-similarity principle were proposed. And an international research centre will be established in IGGE to promote the research works after it will be approved by UNESCO.

Chapter 7 comprises a geochemical study of clays of surficial materials from a Paleozoic and Cenozoic region of SE Iberia. Clay samples were collected for study in the Estremoz-Barrancos sector (SEB) of the Ossa Morena Zone, including the residual clays derived by weathering from diorites and gabbros, granodiorites and tonalites from the Reguengos de Monsaraz massif (RM), from schists, dolerite veins, vulcanites and schists with metabasites, and sedimentary clays (Tertiary), in the Reguengos de Monsaraz –Mourão – Póvoa region. Mineralogical composition obtained by X-ray diffraction of both the bulk and <2 μm fraction able to distinguish some of those clay samples, but it was chemical analyses of trace elements that made an important role to obtain regional geochemical fingerprinting of both residual and sedimentary clays of surficial materials from the Paleozoic and Cenozoic of the studied area. REE patterns enhance the geochemical differences between clay samples, especially by the different degree of fractionation of REE and corresponding slope of REE pattern. The residual clay derived by weathering of the dolerite vein is in general enriched in all elements, especially REE, and on the opposite, the sedimentary Tertiary clays, due to the diluting effect made by the high amounts of carbonates, have the lower amounts of all chemical elements, especially REE, with the exception of As and Sb, which ions are commonly fixed by clays and calcium. The clays from weathering diorites and gabbros, granodiorites and tonalites of RM have similar geochemical behavior as clays from weathered schists with metabasites, with high amounts of elements related with ferromagnesian minerals and HREE; LREE depletion is explained by the presence of heavy minerals like allanite. Clays derived by weathering of schists have geochemical similarities with vulcanite clays, especially due to high amounts of REE, and similar contents of Fe, Sc, Cr, even vulcanites are slightly enriched in Zn, As and Sb. A good geochemical differentiation was achieved within and between residual clay materials and sedimentary clays of surficial materials from the Paleozoic and Cenozoic of the Reguengos de Monsaraz – Mourão - Póvoa region.

In: Advances in Geochemistry Research
Editor: Jorge Sanjurjo-Sánchez
ISBN: 978-1-62618-245-5

Chapter 1

GEOCHEMISTRY OF LOWER PLIOCENE-HOLOCENE FORMATIONS OF DOÑANA NATIONAL PARK (SW SPAIN)

M. Pozo[1,*], M. I. Carretero[2], F. Ruiz[3], M. Olías[3], M. Abad[3], J. Rodríguez Vidal[3], L. M. Cáceres[3], M. I. Prudêncio[4], M. I. Dias[4], E. Font[5], A. M. Muñoz[6], T. Izquierdo[7] and A. Toscano[3]

[1]Departamento de Geología y Geoquímica, Universidad Autónoma de Madrid, Madrid, Spain
[2]Departamento de Cristalografía, Mineralogía y Química Agrícola, Universidad de Sevilla, Sevilla, Spain
[3]Departamento de Geodinámica y Paleontología, Universidad de Huelva, Huelva, Spain
[4]Instituto Tecnologico e Nuclear, Sacavém, Portugal
[5]IDL-Facultade de Ciencias da Universidade de Lisboa, Lisboa, Portugal
[6]Departamento de Biología Ambiental y Salud Pública, Universidad de Huelva, Huelva, Spain
[7]Centro de Vulcanologia e Avaliação de Riscos Geológicos, Universidade dos Açores, Rua da Mãe de Deus, Ponta Delgada, Portugal

[*] Authors for Correspondence: E-mail: manuel.pozo@uam.es.

ABSTRACT

This paper includes the first geochemical analysis of the recently defined formations that conform the Pleistocene-Holocene geological record of Doñana National Park, the most important Biosphere Reserve in Spain. In addition, a multiple geochemical background is proposed by means of the inclusion of unpolluted samples derived from a sediment core collected in the northwestern sector of this area. This background was used to evaluate the impact of the 1998 Aznalcóllar mining spill.

1. INTRODUCTION

In the last decade, numerous studies have focused on the geochemical analysis of estuaries and their potential pollution sources [1]. These researches are especially relevant when the studied estuary contains a high biodiversity [2] and is included in the UNESCO Reserves, National Parks or MAB sites.

In order to clarify the impact of several human-induced activities on soils or sediments in these areas it is interesting to obtain a regional geochemical background. In some cases, values in the standard fossil shale [3] have been used as background (e.g. [4]) while in others, a multiple background was used owing to the high grain size variability of these coastal environments [5].

Doñana National Park (SW Spain) is one of the most important UNESCO Biosphere Reserve of Europe with more than 5000 species. Recently, Carretero et al. (2011) proposed a geochemical background for silty sediments in this area based on the analysis of unpolluted sediments extracted from sediment cores. This paper analyzes the geochemistry of recently described sandy formations of the Park [6] to obtain a geochemical background for these coarser sediments and increase the geochemical background knowledge of Doñana National Park.

2. STUDY AREA

Doñana National Park is one of the largest wetlands in Europe (>50,000 ha), with a very high biodiversity. This area consists of numerous freshwater ponds and freshwater to brackish marshes, partially protected by two spits (Figure 1: Doñana and La Algaida) and some fluvial levees. The main hydrodynamic processes are controlled by the Guadalquivir River discharges (mean: 164 m^3/s), seasonal rains (500-700 mm/year), a mesotidal regime (average tidal range: 3.6 m) and the NE-SW littoral drift currents.

The geological record of the Park has received increasing attention over the last decade, with a recent definition of the main Pleistocene-Holocene formations. The substratum is formed by two Neogene formations (Huelva Sand Formation and Gibraleón Clay Formation; [7]). Above, four Pleistocene-Holocene sedimentary formations (Almonte Sand and Gravel, Lebrija Clay and Gravel, Marisma Clay and Abalario Sand) have been identified [6], interpreted as proximal-alluvial, distal-alluvial, alluvial-estuarine and aeolian deposits, respectively. Sedimentary facies, mineralogy, geochemistry, palaeontology and chronology of some of them have been studied in the last years ([8], [9]), with some palaeogeographical reconstructions of the last 50,000 years [10].

In addition, this record is highly interesting due to the presence of Holocene tsunamigenic deposits (e.g. [11]; [9]) and polluted sediments derived from historical mining activities carried out during the last 4000 years along the Iberian Pyrite Belt, one of the most important metallogenic provinces of Europe [12]. In 1998, the Aznalcóllar disaster released 5 Hm^3 of acid water and metal-rich sludge that gained the external part of the park [13] and highlighted the need for a geochemical background to test the possible impact of human actions.

3. METHODOLOGY

A continuous borehole (Figure 1: Pequeña Holanda-PH-) was drilled in 2007 by the IGME (Geological and Mining Institute of Spain) in the northwestern part of Doñana National Park (UTM: X=719463; Y=4108137). The core is 136.2 m long although only the upper 81 m have been studied in this paper. The first step in this research was the identification of the main formations and facies. Chemical analyses of six selected samples were performed by Activation Laboratory, Ontario (Canada). Element concentrations were determined by Instrumental Neutron Activation Analysis, Inductively Coupled Plasma Spectrometry or MultiINAA-ICP techniques. The calibration is based on the analysis of over 30 international standard reference materials. Detection limits are included in Table 1.

In addition, the palaeontological record of thirteen samples (25 g) was studied, including both macrofauna and microfauna. The total assemblages of both foraminifera and ostracoda were determined at species level.

4. RESULTS AND DISCUSSION

4.1. The Geological Record of Core PH

The lower part of core PH (-81 to -77 m) is represented by bioclastic, grayish fine to very fine sands of the Huelva Sand Formation [7]. The macrofossil record is constituted mainly by fragments and disarticulated valves of ostreids, whereas benthic foraminifera are frequent (*Ammonia tepida*, *Neoconorbina williansonii*, *Heterolepa bellincionii* or several species of *Elphidium*). These features would indicate a deposition in a shallow marine palaeoenvironment, close to the coastline [14].

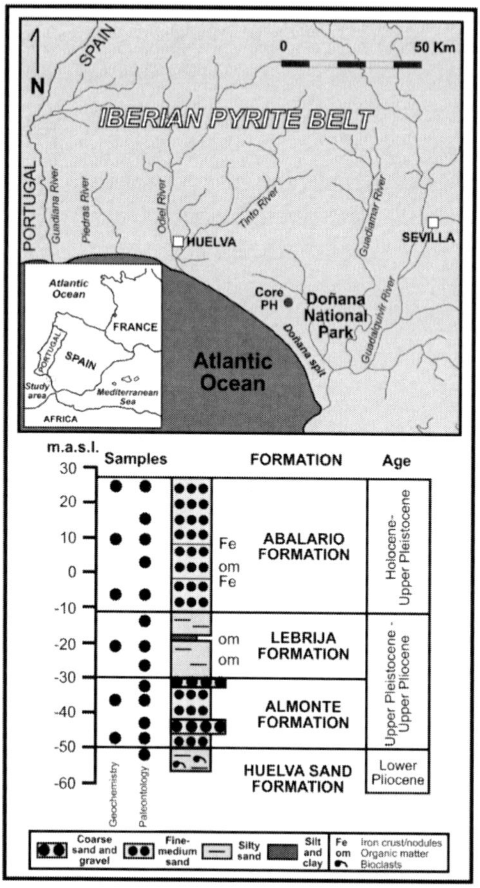

Figure 1. Location and log of the studied core.

The overlying facies (-77 to -54 m: Almonte Formation) is formed by yellowish, quartz-rich fine to coarse sands with variable proportions of gravels. Shell debris (mainly ostreids) is very rare near the base, whereas microfossils (ostracoda, foraminifera) are absent. These sediments present an erosive base and represent the transition to alluvial environments [6].

These deposits are covered by grayish, silty sands and several thick layers of black silty clays with high proportions of organic matter and vegetal remains. They constitute the upper unit of Lebrija Formation and have been interpreted as distal alluvial sediments [6]. The finest, organic-rich sediments include numerous roots, vegetal fragments and utricules of the genus *Carex*, indicating the periodic presence of wetlands, peats, riverbanks or pond margins [15].

The upper 38 m of the core consist by orange to yellowish, fine to medium sands. They are aeolian units with very high proportions of quartz and a lack of fossil remains.Some thick, reddish iron-rich layers are included in these sediments (Abalario Sand Formation; [6]).

4.2. Major Elements and Formations

The geochemical variations of some major elements allow distinguishing the alluvial sediments (Almonte Formation and Lebrija Formation) from the aeolian units (Abalario Formation) and even the proximal-distal origin of the first two formations (Table 1). The proximal-alluvial sediments of Almonte Formation present the highest concentrations of Al (2.47%-2.57%), Mg (0.22%-0.47%), Ca (0.22%-4.42%) or Fe (1.78%-2.01), whereas the distal-alluvial sediments of Lebrija Formation are characterized by moderate values of Na (0.069%), Al (1.49%) or K (1.04%). The high percentages of Ca observed in the lower part of Almonte Formation will indicate a partial marine origin of these sediments, together with the presence of shell debris. This correlation has been observed in other sediment cores obtained in southwestern Spain [16].

These values contrast with the geochemical signature of the Abalario Formation, with low to very low concentrations and a general decrease towards the top.

In these fine to medium sands, the Al content range from 0.47% to 0.82%, whereas Ca or Mg fall in a range 0.02-0.04% and the Fe contents varies between 0.92 and 0.19%. These percentages are very similar to those observed in other dune systems of the Doñana spit [17].

Table 1. Major element geochemistry of selected samples

ELEMENT	Na	Mg	Al	K	Ca	Fe	FORMATION
UNITS	%	%	%	%	%	%	
DETECTION LIMIT	0.001	0.01	0.01	0.01	0.01	0.01	
PH-0.5	0.01	0.02	0.47	0.21	0.02	0.19	Abalario Formation
PH-17	0.016	0.02	0.82	0.5	0.01	0.16	
PH-36	0.041	0.04	1.15	0.69	0.04	0.92	
PH-49	0.069	0.03	1.49	1.04	0.05	0.39	Lebrija Formation
PH-63	0.028	0.25	2.47	0.89	0.22	2.01	Almonte Formation
PH-76	0.103	0.47	2.57	1.4	4.42	1.78	

4.3. Trace Elements

The Almonte Formation presents the highest trace element concentrations of the core (V: 51-52 ppm; Cr: 25.8-27.9 ppm; Ni: 14.6-16.6 ppm; Co: 7.8-16.6 ppm; Zn: 14.3-26.9 ppm; Sr: 28.3-97 ppm; Zr: 57-60 ppm; Ba: 163-204 ppm; Cu: 5.2-15.3 ppm; Pb: 13.8-18.1 ppm). These levels decrease towards the top of the core, with the presence of very low concentrations of these elements near the surface (Table 2: sample PH-0.5). This transition from the basal, coastal sediments of Almonte Formation to the uppermost, alluvial and aeolian palaeoenvironments is well registered in the Sr or Ba values, for example, that suffer an important decrease linked to the restriction of tidal influence.

Table 2. Trace element geochemistry of selected samples

ELEMENT	V	Cr	Ni	Co	Zn	As	Sr	Zr	Ba	Cu	Pb	FORMATION
UNITS	ppm	ppm	ppm	ppm	ppm	ppm	ppm	ppm	ppm	ppm	ppm	
DETECTION LIMIT	1	0.5	0.5	0.1	0.2	0.1	0.2	1	1	0.2	0.5	
PH-0.5	6	3.8	1.9	0.6	< 0.2	1.1	6.1	2	32	2.4	4.7	Abalario Formation
PH-17	5	10.9	1.5	0.7	< 0.2	< 0.1	9.8	3	80	2.1	3.9	
PH-36	8	12.6	2.2	1.5	0.7	2.8	16.6	11	113	1.4	6.7	
PH-49	6	5.9	1.9	0.9	< 0.2	0.4	22.7	9	171	2.6	7.4	Lebrija Formation
PH-63	52	25.8	14.6	16.6	14.3	1.3	28.3	60	163	5.2	18.1	Almonte Formation
PH-76	51	27.9	16.6	7.8	26.9	3.7	97	57	204	15.3	13.8	

These elements have been used as markers of marine inputs in numerous investigations of coastal cores, because they occur in higher concentrations in seawater than freshwater [18]. A similar trend was observed for Zr in some cases [19]. Consequently, core PH represents a general regressive sequence from the basal, intertidal sediments to the uppermost, aeolian deposits.

4.4. Regional Geochemical Background

Our data and those extracted from clayey, unpolluted sediments of Lebrija Formation (samples PLN-5 to PLN-9; [12]) can be used to obtain a regional geochemical background (Table 3). We propose to use the uppermost sample of core PH (PH-0.5) as the background for aeolian, sandy sediments and PH-49 sample as background for silty sands.

The 1998 Aznalcóllar mining spill can now be evaluated. This event caused a variable pollution of some sandy areas located very close to the Doñana National Park [20], with moderate to high contents of Pb (44.7-86.7 ppm; 9-18.5 BK) or Cu (15.9-53.3 ppm; 6.5-22 BK).

Table 3. Proposed geochemical background (BK) for Doñana National Park

GRAIN SIZE	ENVIRONMENT	As	Co	Cr	Cu	Pb	Ni	REFERENCE
Sand	Aeolian	1.1	0.6	3.8	2.4	4.7	1.9	This paper
Silty sand	Alluvial	0.4	0.9	5.9	2.6	7.4	1.9	This paper
Silt and clay	Estuarine-alluvial	14	11.2	69.6	19	12.6	27.2	Carretero et al. (2011)

CONCLUSION

The multidisciplinary analysis of a long core collected in the northwestern part of Doñana National Park allows inferring a temporal evolution from Pliocene, infralittoral areas to Holocene, aeolian environments, with an Pleistocene alluvial interlude. This core contains unpolluted sediments of different grain sizes used to propose a regional background. A first application of this background to the 1998 Aznalcóllar mining spill indicates a moderate to high pollution (6.5-22 BK) for different metals (e.g. Pb, Cu) in adjacent, sandy areas.

Acknowledgments

This work was funded by a Spanish DGYCIT Project (CGL2010-15810/BTE), an Andalusia Excellence Project (SEJ-4770: Ordenación y gestión integradas del litoral de Andalucía) granted by FEDER, a Project funded by the Fundacão para a Ciência e a Tecnologia of Portugal (PTDC/CTE-GIX/110205/2009) and three Research Groups of the Andalusia Board (RNM-238, RNM-293 and RNM-349). This paper is a contribution to the IGCP 588 (Preparing for coastal change).

References

[1] Castro, P., Valiela, I. and Freitas, H. (2007) The use of sedimentary %C, %N, $\delta^{15}N$, and Pb concentrations to assess historical changes in anthropogenic influence on Portuguese estuaries. *Environ. Pollut.*, 147, 706–712.

[2] Taggart, M. A., Figuerola, J., Green, A. J., Mateo, R., Deacon, C., Osborn, D. and Meharg, A. A. (2006) After the Aznalcóllar mine spill: Arsenic, zinc, selenium, lead and copper levels in the livers and bones of five waterfowl species. *Environ. Res.*, 100, 349–361.

[3] Turekian, K. K. and Wedepohl, K. H. (1961). Distribution of the elements in some major units of the earth crust. *Bull. Geol. Soc. America*, 72, 175-192.

[4] Magesh, N. S., Chandrasekar, N. and Vetha Roy, D. (2011). Spatian analysis of trace element contamination in sediments of Tamiraparani estuary, southeast coast of India. *Est. Coast Shelf Sci.*, 92, 618-628.

[5] Ruiz, F. (2001). Trace metals in estuarine sediments from the southwestern Spanish coast. *Mar. Poll. Bull.*, 42, 482-490.

[6] Salvany, J. M., Cruz, J., Mediavilla, C. and Rebollo, A. (2011). Chronology and tectono-sedimentary evolution of the Upper Pliocene to Quaternary deposits of the lower Guadalquivir foreland basin, SW Spain. *Sed. Geol.*, 241, 22-39.

[7] Civis, J., Sierro, F. J., González Delgado, J. A., Flores, J. A., and Valle, M. F. (1987). El Neógeno marino de la provincia de Huelva: Antecedentes y definición de las unidades litoestratigráficas. In: *Paleontología del Neógeno de Huelva (W Cuenca del Guadalquivir)* (Civis, J., ed.), 9-23. Universidad de Salamanca, Spain.

[8] Ruiz, F., Abad, M., Rodríguez Vidal, J., Cáceres, L. M., González-Regalado, M. L., Carretero, M. I., Pozo, M. and Gómez Toscano, F. (2008). The geological record of the oldest historical tsunamis in southwestern Spain. *Riv. Ital. Paleontol. Strat.*, 114, 147-156.

[9] Pozo, M., Ruiz, F., Carretero, M. I., Rodríguez Vidal, J., Cáceres, L. M., Abad, M. and González-Regalado, M. L. (2010). Mineralogical assemblages, geochemistry and fossil associations of Pleistocene-Holocene complex siliciclastic deposits from the Southwestern Doñana National Park (SW Spain): A palaeoenvironmental approach. *Sed. Geol.*, 225, 1-18.

[10] Ruiz, F., Pozo, M., Carretero, M. I., Abad de los Santos, M., González-Regalado, M. L., Muñoz Pichardo, Rodríguez Vidal, J., Cáceres, Puro, L. M., Pendón, J. G., Prudêncio, M. I. and Dias, M. I. (2010). Birth, Evolution and Death of a Lagoon: Late Pleistocene to Holocene Palaeoenvironmental Reconstruction of the Doñana National Park (SW Spain). In: *Lagoons: Biology, Management and Environmental Impact* (Friedmann, A. G., ed.), 371-396. Nova Science Publishers, New York.

[11] Ruiz, F., Rodríguez-Ramírez, A., Cáceres, L. M., Rodríguez Vidal, J., Carretero, M. I., Abad, M., Olías, M. and Pozo, M. (2005). Evidence of high-energy events in the geological record: Mid-Holocene evolution of the southwestern Doñana National Park (SW Spain). *Palaeogeog., Palaeoclimatol., Palaeoecol.*, 229, 212-229.

[12] Carretero, M. I., Pozo, M., Ruiz, F., Rodríguez Vidal, J., Cáceres, L. M., Abad, M., Muñoz, J. M., Gómez, F., Campos, J. M., González-Regalado, M. L., and Olías, M. (2011). Trace elements in Holocene sediments of the southern Doñana National Park (SW Spain): historical pollution and applications. *Environ. Earth Sci.*, 64, 1215-1223.

[13] Meharg, A. A., Osborn, D., Pain, D. J., Sánchez, A. and Naveso, M. A. (1999). Contamination of Doñana food-chains after the Aznalcóllar mine disaster. *Environ. Pollut.*, 105, 387-390.

[14] González-Regalado, M. L. (1989). Estudio sistemático de los foraminíferos bentónicos de las arenas fosilíferas del Plioceno de Huelva: su significado paleontológico. *Est. Geol.*, 45, 101-119.

[15] Klötzi, F. (1988). Conservational status and use of sedge wetlands. *Aquatic Botany*, 30, 157-168.

[16] Ruiz, F., Borrego, J., López-González, N., Abad, M., González-Regalado, M. L., Carro, B., Pendón, J. G., Cáceres, L. M., Prudêncio, M. I. and Dias, M. I. (2007). The geological record of a mid-Holocene marine storm in Southwestern Spain. *Geobios*, 40, 689-699.

[17] Abad, M., Ruiz, F., Pozo, M., Carretero, M. I., Rodríguez Vidal, J., Cáceres, L. M., Muñoz, J. M., González-Regalado, M. L., Pendón, J. G.,Toscano, A., Borrego, J., Prudencio, M. I. and Dias, M. I. (2011). Correlation between millennial mining activities and heavy metal pollution in two Southwestern Spanish estuaries. In: *Heavy metal sediments* (Sanz, L. H, Ed.), 103-114. Nova Science Publishers, New York.

[18] Ramírez-Herrera, M. T., Lagos, M., Hutchinson, I., Kostoglodov, V., Machain, M. L., Caballero, M., Goguitchaichvili, A., Aguilar, B., Chagué-Goff, C., Goff, J., Ruiz Fernández, A. C., Ortiz, M., Nava, H., Bautista, F., López, G. and Quintana, P. (2012). Extreme wave deposits on the Pacific coast of Mexico: Tsunamis or storms?-A multiproxy approach. *Geomorphology*, 139-140, 360-371.

[19] Goff, J., Pearce, S., Nichol, S.L., Chagué-Goff, C., Horrocks, M. and Strotz, L. (2010). Multiproxy records of regionally-sourced tsunamis, New Zealand. *Geomorphology*, 118, 369–382.

[20] Riba, I., Conradi, M., Forja, J. M. and DelValls, T. A. (2004). Sediment quality in the Guadalquivir estuary: lethal effects associated with the Aznalcóllar mining spill. *Mar. Poll. Bull.*, 48, 144-152.

In: Advances in Geochemistry Research
Editor: Jorge Sanjurjo-Sánchez
ISBN: 978-1-62618-245-5

Chapter 2

GEOCHEMISTRY AND MINERALOGY OF PLIOCENE CLAY DEPOSITS FROM WEST PORTUGAL: PROVENANCE SIGNALS?

***Álvaro Oliveira*[1], *Pedro A. Dinis*[2,*] *and Fernando T. Rocha*[3]**

[1]Laboratório do INETI de S. Mamede Infesta, Porto, Portugal
[2]Department of Earth Sciences, IMAR-CMA, University of Coimbra, Portugal
[3]Geobiotec Centre, Department of Geosciences, University of Aveiro, Portugal

ABSTRACT

In this work is presented a comparative study of mineralogical and geochemical (major elements) data of Pliocene clay deposits from west Portugal. Quartz occurs in larger amounts in the coarser sediments and is usually the most common mineral in the deposits north of Mondego river. Illite proportion is always higher in the basins to the north of Mondego, while kaolinite tends to be higher in the southern basins, where it is often the most common mineral. One set of samples from lower stratigraphic position of the northern sector (Vale Grande) has mineralogical features that resemble the southern deposits. In accordance with the mineralogical

* Corresponding author. E-mail: pdinis@dct.uc.pt.

composition, silica and alumina are the most common oxides. Silica tends to be higher in coarser samples, in particular in the deposits south of Mondego river. Potassium oxide content is always higher in the northern than in the southern deposits. The observed differentiation between clay deposits is attributed to provenance. The deposits located to the north of Mondego river were derived mainly from the Precambrian metapelites of the Central Iberian Zone. Regarding the deposits to the south of Mondego river, the dominance of kaolinite and high Al_2O_3 content suggest that they result mainly from recycling of Cretaceous and Cenozoic units, which, in turn, derive from weathering and erosion of granitoids and metasedimentary rocks. Some older deposits north of Mondego were also originally sourced by previous cycle sediments, but provenance must have shifted to basin-edge metapelitic rocks due to uplifting of coastal mountains, which blocked the drainage from the hinterland. Besides provenance, hydraulic selection could explain local compositional differences.

INTRODUCTION

The composition of sedimentary units is conditioned by the intensity of chemical weathering before deposition, which is likely to be promoted when source areas are characterized by wetter and warmer climate and low relief [1, 2]. Other factors during a sediment cycle, like provenance and sorting processes, may also play a major role in determining the composition of a sedimentary unit [3, 4, 5, 6, 7, 8, 9, 10, 11, 12]. Furthermore, post-depositional changes are also able to shift the composition of a sedimentary unit [3, 13, 14, 15].

In Portugal, the composition of sedimentary deposits has proved to be very useful in establishing lateral equivalences of stratigraphic successions [16, 17, 18]. To do so, it is generally assumed that deposits with features pointing to intense chemical weathering (e.g. high kaolinite content) are formed during periods of wetter and warmer climates. However, compositional features attained after intense chemical alteration are persistent at surface [3, 13]. Hence, even if climatic conditions coeval of a sedimentary deposits had become less chemically aggressive, the clay mineral assemblages and geochemistry may reflect previous stages, when weathering was more intense.

In this work we compare the mineralogy and geochemistry (major elements) of several Pliocene west-Portuguese clay deposits that are considered stratigraphic equivalents. Later, we discuss the factors that may explain the differences found between these deposits.

GEOLOGICAL SETTING

Two sectors from west Portugal, south and north of Mondego river, approximately 40 km apart, were selected for this study. The Pliocene south of Mondego river integrates a lower portion, which is mainly sandy in lithology, and an upper portion, which include thick mud deposits [19, 20, 21]. The sandy portion starts with well sorted, fine to very fine sand, frequently micaceous, sometimes with lens-shaped pebble beds. The sand portion evolves upward to coarser deposits, usually white or light grey. These units are referred to by Carnide Sandstones and Roussa Sandstones, respectively [20]. The transition to the upper portion (Barracão Mudstones) is usually marked by a mud bed with variable thickness, sometimes intercalated with lignite. Besides the fine-grained mud deposits, this formation also include poorly sorted coarse sand and gravel deposits that become progressively more common upwards.

The deposits north of the Mondego river are located along the contact zone between the Atlantic margin, to the west, and the Variscan Massif, to the east, which is associated with the Porto-Tomar Fault Zone (Figure 1). Within this sector are observed several faults responsible for the development of subsiding areas where the Pliocene and Quaternary sedimentary succession is usually especially thick (up to 70m) [22, 23, 24]. The upper Neogene succession that precedes Quaternary fluvial incision was divided in four sedimentary units, defining a regressive sequence [24]. Unit U1 starts by yellow to grey color, fine to very fine sand, usually well sorted, that sometimes evolves upward to coarser and less sorted sediments. Unit U2B covers by unconformity U1 or the basement. It is formed by gravel and gravelly-sand, with rare and thin (<0.5 m) mud beds. Unit U2B evolves gradually upward from U2A or constitutes its lateral equivalent in the most subsidence sectors. The sequence is truncated and ends up by the gravelly beds of unit U3.

Materials and Methods

The studied samples were collected from exploration boreholes drilled in the regions to the north (28 samples) and to the south of Mondego river (24 samples) (Table 1). The samples from the basins north of Mondego river were collected in the unit U2B and came from Vale Grande (5 samples), Aguada (7 samples), Anadia (5 samples) and Monsarros (12 samples) areas (Figure 1). The samples from the basins south of Mondego river were collected in Barracão Formation and came from Pega (7 samples), Redinha (9 samples) and Meirinhas (8 samples) areas. Nine samples from previous Cretaceous-Cenozoic sedimentary units were analyzed for means of comparison.

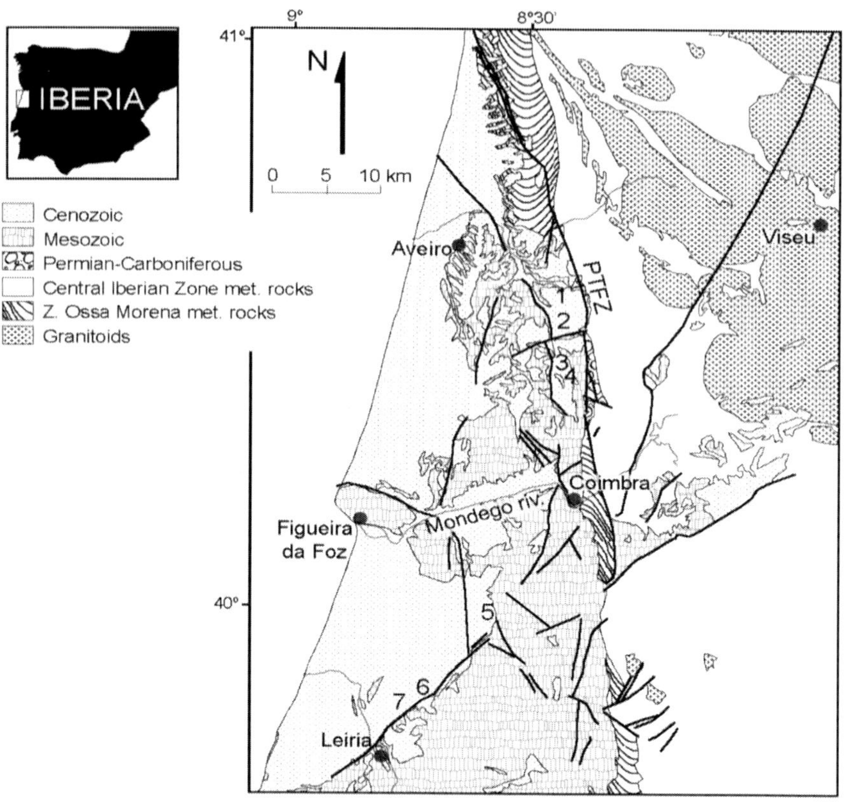

Figure 1. Geological framework and location of the studied deposits.
1: Vale Grande; 2: Aguada; 3: Anadia; 4: Monsarros; 5: Redinha; 6: Meirinhas; 7: Pega. PTFZ: Porto-Tomar fault zone.

Table 1. Mineralogy and geochemistry of the studied Pliocene samples

		Qtz	Il	Kaol	Fk	Rut	Hem	Gt	Il*	Kaol*	Qtz*	SiO2	Al2O3	Fe total	MnO	CaO	MgO	Na2O	K2O	TiO2	P2O5	P.R.
Vale Grande	8291NM	30	31	34	2		2		16	80		60,13	21,85	5,49	0,02	0,04	0,34	0,27	2,29	1,18	0,04	8
	8293NM	14	49	35	1		1		16	76	1	61,35	23,24	3,27	0,02	0,05	0,4	0,29	3,01	0,82	0,07	7,23
	8294NM	34	30	32	2		2		25	72	1	61,82	21,66	3,94	<0,02	0,07	0,44	0,27	2,18	0,85	0,07	8,34
	8300NM	40	37	20	1				20	77	vest.	60,65	22,19	4,9	<0,02	0,08	0,4	0,29	2,53	0,71	0,09	7,78
	8303NM	40	29	27	2				14	82	1	65,03	19,26	4,56	0,02	0,02	0,3	0,25	1,79	0,91	0,04	7,5
Aguada	7194NM	34	45	16	3		2		46	29	17	64,9	18,2	5,91	0,02	0,04	0,42	0,20	3,27	0,88		5,05
	7197NM	45	33	16	6				48	28	20	66,08	21,14	2,11	0,02	0,03	0,34	0,20	3,17	1,1		5,52
	7199NM	44	28	21	6		1					64,72	20,27	3,59	0,02		0,41	0,20	2,86	1,19	0,03	6,1
	7162NM	25	40	28	2	3						52,5	28,22	1,67	0,02	0,21	0,66	0,20	3,22	0,56	0,04	12,45
	7196NM	32	42	23	1	1						60,18	25	1,89	0,02	0,04	0,476	0,20	4	1,04		6,61
Anadia	7024NM	51	25	12	5	3	4	0	46	42	8	64,26	17,6	7,04	0,02	0,04	0,43	0,20	2,68	0,93	0,05	6,63
	7028NM	41	32	18	3	1	3	2	46	47	3	65,5	18,9	4,59	0,02	0,04	0,47	0,20	2,87	0,93	0,03	6,27
	7031NM	35	39	20	2	1	3	1	41	41	8	62,49	20,87	3,91	0,02	0,05	0,42	0,20	2,46	1,11	0,03	8,32
	7044NM	47	31	18	3	0	0	1	49	43	5	65,9	18,6	4,53	0,02	0,06	0,48	0,20	2,94	0,98	0,05	6,12
	7048NM	35	33	24	7	0	1	0	55	39	4	62,43	21,61	1,87	0,02	0,1	0,88	1	3,29	0,9	0,06	7,71
Monsarros	7136NM	51	25	17	5	1	-	1	45	43	7	64,34	21,00	2,08	0,02	0,04	0,34	0,2	2,76	1,07	0,04	8,27
	7147NM	54	26	11	3	1	4	1	54	36	5	71,2	15,46	4,49	0,02	0,04	0,31	0,2	2,59	0,85	0,04	4,83
	7149NM	39	31	16	5	1	5	3	54	38	4	64,02	20,32	4,2	0,02	0,04	0,39	0,2	3,28	1,08	0,05	6,46
	7152NM	47	24	15	6	2	2	1	60	32	5	62,72	22,21	3,02	0,02	0,06	0,44	0,2	2,93	1,15	0,03	7,14
	7159NM	49	30	14	4	1	2	-	61	29	5	66,52	20,05	1,89	0,02	0,08	0,49	0,2	3,3	1	0,04	6,54
	7160NM	54	23	14	4	1	3	1	48	40	5	68,88	16,36	4,47	0,02	0,08	0,4	0,2	2,71	0,92	0,07	5,88
	8498NM	40	32	17	4	1	6		52	39	5	63,55	17,77	7,42	<0,02	0,04	0,36		2,83	1,02	0,11	6,35
	8499NM	27	42	24	4	1	2		47	42	6	60,68	22,88	3,35	<0,02	0,06	0,5	0,24	3,32	1,14	0,05	7,46
	8501NM	40	36	17	3	1	3		54	40	4	69,06	16,56	4,07	<0,02	0,05	0,36	0,28	2,89	0,96	0,06	5,38
	8502NM	33	42	20	3	1	1		55	40	3	67,33	19,17	1,79	<0,02	0,06	0,47	0,26	3,23	1,1	0,04	6,2
	8504NM	37	37	19	4	2	1		49	40	7	65,13	20,77	0,99	<0,02	0,15	0,53		3,33	1,17	0,04	7,61
	8505NM	44	25	16	9	1	5		46	42	8	64,97	18,41	5,16	0,02	0,11	0,44		2,85	0,98	0,12	6,51

Table 1. (Continued)

		Qtz	Il	Kaol	Fk	Rut	Hem	Gt	Il*	Kaol*	Qtz*	SiO2	Al2O3	Fe total	MnO	CaO	MgO	Na2O	K2O	TiO2	P2O5	P.R.
Redinha	6068NM	8	13	73	5	1		vest	9	89		48,58	30,46	1,24	0,02	0,08	0,23	0,2	0,52	0,71	0,05	17,82
	6075NM	14	13	55	1		10	2	13	78	2	45,2	27,67	10,57	0,02	0,17	0,52	0,08	0,76	1,07	0,32	13,45
	6078NM	13	10	75	1			vest	11	85	1	48,63	32,72	2,27	0,02	0,18	0,43	0,2	0,72	0,75	0,12	13,83
	6055NM	90	0	10	1		0		14	81	5	84,9	8,78	0,84	0,02	0,04	0,21	0,2	0,39	0,19	0,03	4,11
	6056NM	18	13	68	1		1		15	83	2	54,44	28,19	2,16	0,02	0,14	0,64	0,2	1,46	0,75	0,08	11,77
	6061NM	42	1	56	2		2		7	93	0	62,45	21,17	4,69	0,02	0,11	0,39	0,2	0,79	1,2	0,11	8,67
	6064NM	20	1	73	4		7		19	80	1	47,56	27,86	8,44	0,02	0,13	0,92	0,2	1,58	1,21	0,07	11,86
	6073NM	32	8	57	3		3		11	88	1	55,09	26,645	3,21	0,02	0,13	0,6	0,2	1,26	1,31	0,04	11,47
	6076NM	76	1	24	1		0		12	88	0	71,5	16,87	1,32	0,02	0,11	0,51	0,2	0,81	1,02	0,12	7,23
Meirinhas	7783NM	55	7	37	1				19	80	1	76,63	14,25	1,12	< 0,02	0,07	0,23	0,2	0,69	0,53	0,03	6,01
	7790NM	35	10	54	1				20	78	2	72,82	16,45	2,19	< 0,02	0,09	0,24	0,2	1	0,64	0,04	6,28
	7791NM	33	9	57	1							62,55	22,42	3,26	< 0.02	0,11	0,33	0,2	1,05	0,77	0,03	9,47
	7797NM	35	18	47	vest				18	80	2	73,93	15,53	1,93	< 0,02	0,06	0,26	0,2	1,09	0,72	0,05	5,78
	7809NM	44	18	36	vest				21	77	2	78,82	12,89	0,86	< 0,02	0,07	0,26	0,2	0,99	0,68	0,04	5,04
	7807NM	26	18	56	0							59,76	25,45	1,85	< 0.02	0,13	0,47	0,2	1,47	0,74	0,05	9,82
	7811NM	38	16	45	1							64,1	19,79	1,95	< 0.02	0,09	0,43	0,2	1,26	0,54	0,04	11,45
Pega	5979NM	9	8	64	vest			vest.	8	64	9	53,18	27,42	2,64	0,02	0,22	1,08	0,23	1,56	0,85	0,06	12,46
	5981NM	21	13	57					10	78	2	56,06	26,23	2,98	0,02	0,04	0,72	0,15	1,58	0,78	0,04	11,2
	5989NM	22	9	67	0.5		0	0	10	89	1	62,47	23,04	2,46	0,02	0,06	0,48	0,2	1,12	0,95	0,07	9,05
	5990NM	30	11	57	2		0	0	16	82	2	59,23	23,46	2,88	0,02	0,14	0,73	0,2	1,32	0,84	0,05	10,97
	5994NM	9	8	78	3		0.5	2	11	77	3	51,29	28,3	2,99	0,02	0,21	1,07	0,2	1,49	0,8	0,06	13,63
	5996NM	51	0.5	45	3		2	0	12	85	1	73,17	14,78	3,06	0,02	0,04	0,28	0,2	0,74	0,77	0,03	6,75
	5997NM	10	8	77	3		3	0	11	87	2	51,89	27,81	4,26	0,02	0,5	0,48	0,2	1,42	0,8	0,05	12,85
	6002NM	8	10	77	vest				13	87	vest	51,96	29,95	2,97	0,02	0,08	0,44	0,2	1,72	0,64	0,06	11,75

Qtz: Quartz; Il: Illite; Kaol: Kaolinite; Fk: K-feldspar; Rut: Rutile; Hem: Hematite; Gt: Goethite; *: Results for fraction < 2μm.

The mineralogical analysis was performed by X-ray diffraction (XRD). An evaluation of mineral proportions was obtained from distinctive XRD peak areas weighted by empirical factors [25]. Given the uncertainties of the adopted methodology, the obtained results are semi-quantitative approximations of actual mineral proportions [26, 27]. The uncertainties are expected to be higher in the bulk sample XRD estimations, which may involve several interference peaks. Chemical composition (major elements) was determined by X-ray fluorescence (XRF) after digestion of homogenized powders of the samples with a mixture of HF, HNO_3 and $HClO_4$. The laboratory precision for all XRF determinations considered in this work is above 5%. Grain-size was determined by sieving (>350 μm) and laser diffraction (< 350 μm). The proportions of grain-size fractions (< 4 μm, 4-63 μm, 63-2000 μm and > 2000 μm) were determined from combined sieving-laser grain-size distribution curves.

Clay Deposits Mineralogy and Geochemistry

The mineralogy of the Pliocene sediments is characterized by an association of quartz, illite and kaolinite, all in variable amounts (Figure 2A). In smaller proportions occur feldspar and goethite, and, occasionally, smectite, hematite and rutile. Quartz proportions are highly variable and tend to be higher in coarser samples. The amounts of illite are always higher in the clay deposits north of Mondego river (~ 23-49%) than in the clay deposits south of Mondego river (~ 0-18%). The kaolinite proportions, in turn, tend to be much higher in the southern deposits (~ 10-78%), where it is often the most common mineral, than in the northern deposits (~ 11-35%). Vale Grande samples are characterized by higher kaolinite content (~ 20-35 %) than the other northern samples (~ 11-24%).

Similar mineralogical trends are observed in the clay fraction (< 2 μm). Vale Grande samples are enriched in kaolinite (72-82 %) and are plotted together with the southern deposits (Figure 2B). If Vale Grande samples are not considered, illite content is substantially higher in the northern deposits (~ 41-61 %) than in the southern deposits (~ 7-21 %); whilst kaolinite is higher in the southern deposits (~ 64-93 %) than in the northern deposits (~ 28-47 %) (Figure 2B). Smectite, vermiculite, mixed layer clays, hematite, gohetite and quartz are common accessory minerals in the clay fraction. Their summed proportions are always limited (<28 %; usually < 10%).

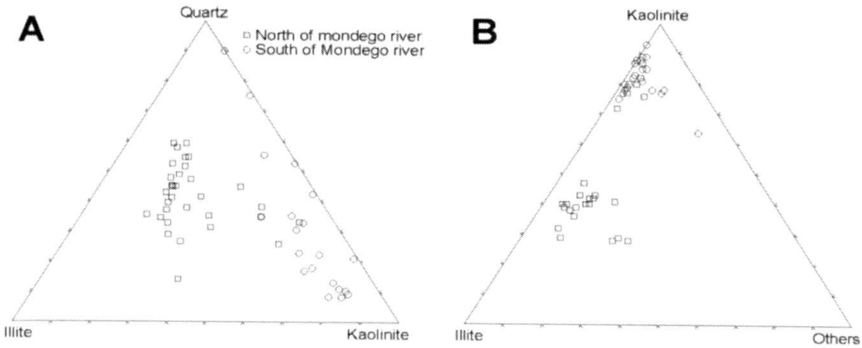

Figure 2. Mineralogy of the Pliocene deposits. Bulk sample (A) and clay fraction (B).

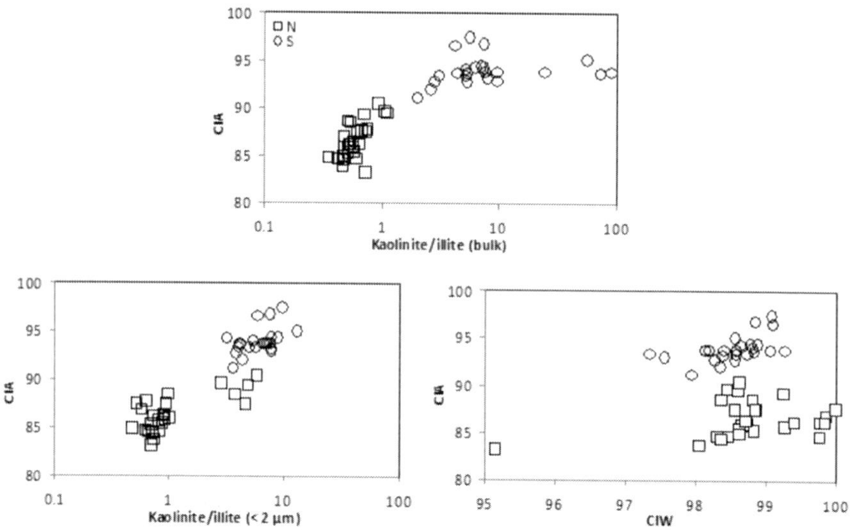

Figure 3. Discrimination of clay deposits based on mineralogical and geochemical features that reflect chemical weathering. The cluster of samples from the northern sector plotted together with southern deposits in the diagram kaolinite/illite (2µm)-CIA is from Vale Grande. N: Clay deposits north of Mondego river: S: Clay deposits south of Mondego river.

In regards to chemical composition, silica (45.2-84.9 %) and alumina (12.7-32.7 %) are the most common oxides. The proportions of these oxides are particularly heterogeneous in the southern clay deposits (45.2-84.9 % SiO_2; 8.78-32.78 % Al_2O_3) and more homogenous in the northern deposits (52.5-71.2 % SiO_2; 15.46-28.22% Al_2O_3). The coarser sediments, namely the deposits from Meirinhas and one sample from Pega, in the southern sector, have the highest silica content. The coarse grained samples from Vale Grande do not have higher silica contents than the other northern samples. The samples with highest alumina contents are from fine-grained beds from Aguada (northern sector), Redinha and Pega (southern sector).

In minor amounts, but still largely variable, occurs $FeO_{(tot)}$ (0.84-10.57 %) and K_2O (0.39-4 %). Except for two fine-grained sample enriched in iron (and also in hematite), the proportion of $FeO_{(tot)}$ tend to be lower and more homogeneous in the deposits south of Mondego (0.84-4.69 %) than north of this river (0.99-7.42 %). The content of K_2O are always higher in the northern deposits (> 1.79 %) than in the southern deposits (< 1.72 %). Except for some samples, the proportions of all the other oxides are lesser than 1 %, being TiO_2 (0.19-1.31%) and MgO (0.21-1.08 %) the most frequent.

CONTROLS ON SEDIMENT COMPOSITION

Chemical Weathering

When plotting chemical and mineralogical compositional data that are influenced by chemical alterations, the sectors in the north and south of Mondego river are clearly separated (Figure 3). Given the enrichments in illite and kaolinite in northern and southern deposits respectively, the relative amounts of these minerals and the ratio kaolinite/illite, by themselves, allow a good separation of the two sectors. Regarding geochemistry, several indices have been proposed to evaluate the intensity of chemical weathering [28]. One of the most commonly used is the ―Chemical Index of Alteration" (CIA) defined by Nesbitt and Young [1]:

$$CIA = 100xAl_2O_3/(Al_2O_3+CaO^*+Na_2O+K_2O)$$

The CIA values are also always higher in the southern sector (91-97) than in the northern sector (83-89). These results are consistent with what we already referred for the mineralogical association, reflecting the greater relative proportions of kaolinite in the southern region and illite in the northern region.

Another index commonly used is the ―Chemical Index of Weathering" (CIW) of Harnois [29]:

$$CIW = 100xAl_2O_3/(Al_2O_3+CaO^*+Na_2O)$$

Because CIW does not consider K_2O contents, this index is found to be ineffective in the discriminations of the sectors under analysis (Figure 3). In fact, the differences between the two regions in terms of compositional maturity appear to be associated with the proportions of K_2O (Figure 4). This oxide reveals significant (probability of 95%) positive correlation with the proportion of illite and negatively correlation with CIA. The correlation K_2O-illite suggests that K_2O is supported mainly by illite, although other minerals, like K-feldspar, could also retain a significant proportion of this oxide, in particular in coarser samples.

The two sets of sampled units have been considered lateral equivalents [24, 30] and it is expected that any slight differences in latitude (the basins are just approximately 40 km apart) and elevation (current elevations of the deposits range between 50 and 150 meters and they were probably formed in a coastal and alluvial plain of low altitude) are not sufficient to justify sound differences in climatic conditions. Hence, the observed compositional diversity should not be related to distinct climate or post-depositional modifications. The possible role of provenance and sediment sorting on the studied deposits composition is discussed below.

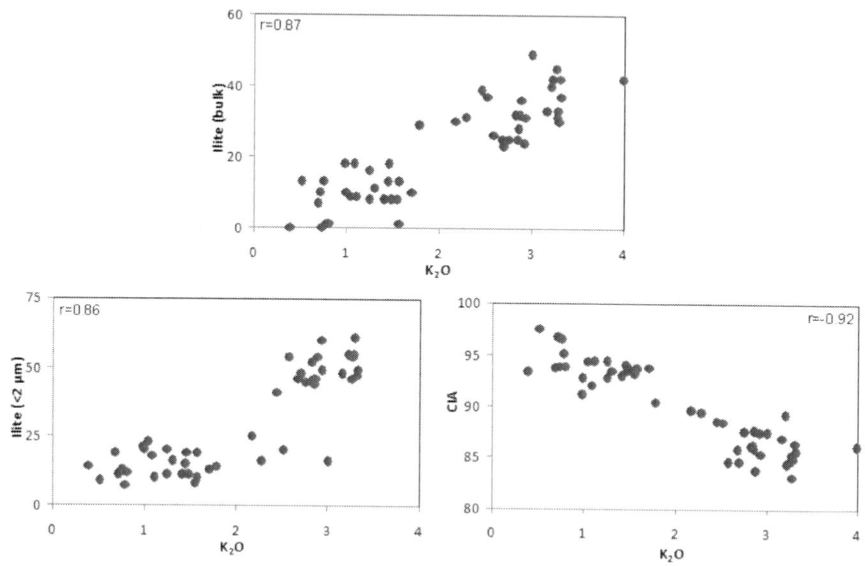

Figure 4. Relation between K_2O contents and the proportions of illite (bulk sample and fraction < 2μm) and CIA values.

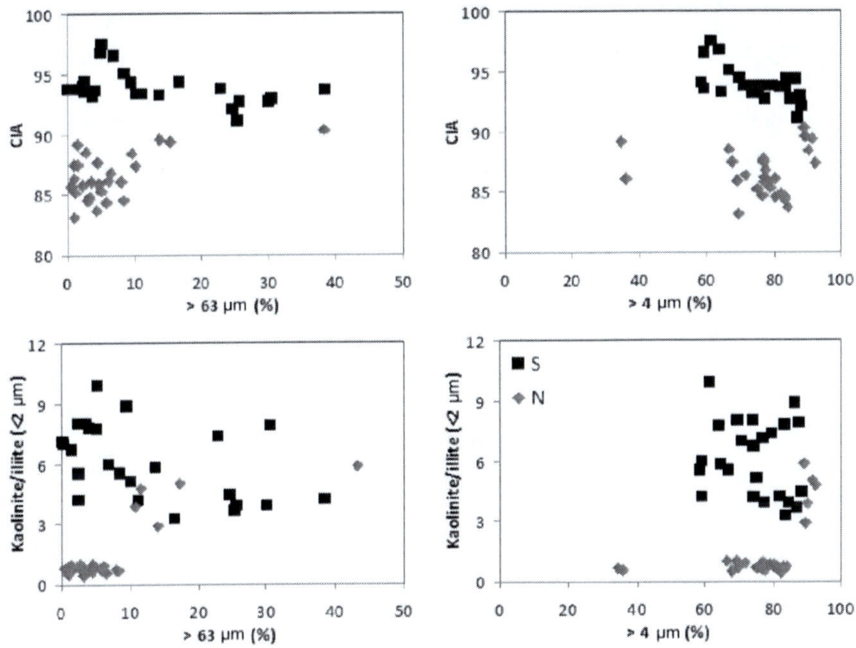

Figure 5. Relation between grain-size parameters and compositional features that reflect the extent of chemical weathering. N: Clay deposits north of Mondego river; S: Clay deposits south of Mondego river.

Sediment Sorting

Sorting processes during a sedimentary cycle are responsible for segregation of different particles according to size, shape and density, which can influence the composition of sedimentary units [5, 6, 9, 31, 32]. The grain-size variability among the studied set of samples must be responsible for some compositional differences. Higher percentages of quartz and silica are usually observed in coarser sediments and alumina and phylossilicates in finer sediments. The coarser grained samples from Vale Grande area tend to have substantially higher kaolinite/illite ratio than the other sediments from the northern sector, even though CIA values are comparable (Figure 3).

From the plots grain-size vs. compositional data it is possible to distinguish several groups of sediments (Figure 4). The best differentiation is found in the plot clay fraction (< 4 µm) proportion – CIA. Vale Grande samples are grouped in a cluster isolated from the other northern samples and seem to define, along with samples from the southern deposits, a tendency for higher chemical weathering in sediments with higher quantities of clay fraction. Higher proportion of clay minerals in some samples may reflect an association with lower relief catchment areas, where chemical weathering was also probably accomplished more easily. It must be mentioned that alteration through exogenous processes is also likely to occur in other phases of a sediment cycle, namely during alluvial storage after particles erosion and transport from inland catchment areas.

Assuming similar provenance and climatic conditions, samples with higher chemical maturity and higher proportion of clay fraction could result from combined effects of higher chemical weathering and lower dynamic conditions in a low relief environment. Under these circumstances, sand and coarse silt fractions, which tend to be enriched in feldspar and diverse heavy minerals, even in environments where chemical weathering is limited [33], must have been deposited in upstream locations.

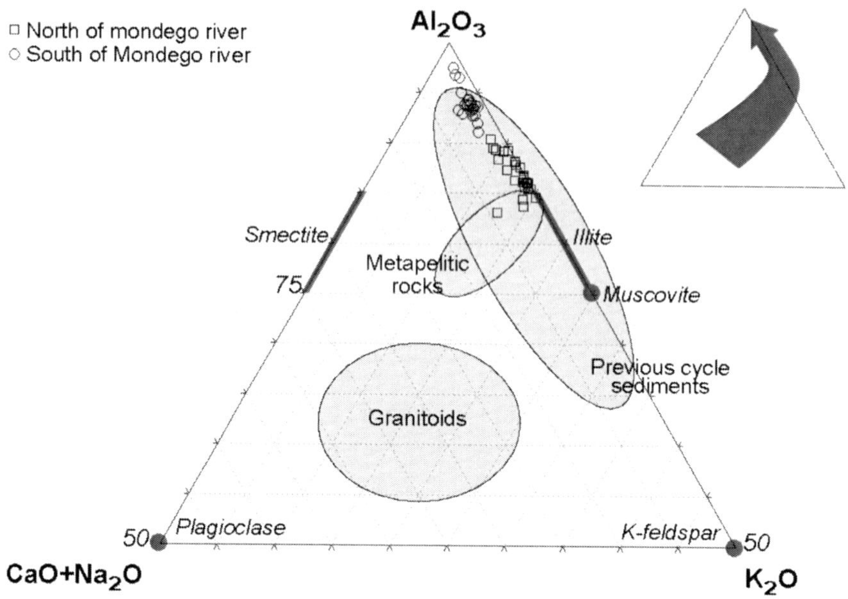

Figure 6. Ternary plot (more than 50% Al2O3) of molecular proportions Al2O3–(Na2O+CaO*)–K2O for the Pleistocene clay deposits. Grey patches indicate the range of values observed in previous cycle sedimentary units, granitoids from Viseu region [43] and metapelitic basement rocks [34]. The arrow in small ternary diagram represents the expected evolution in chemical composition due to progressive weathering.

Provenance

The largest amounts of illite (Figure 2) and K_2O (Figure 6) in the Pliocene units north of Mondego river are most likely related to specific sediment provenance. Actually, along this sector the flysch-like slates and greywackes of the Paleozoic and Precambrian Variscan Iberian Massif, which are enriched in micaceous minerals, are the most plausible source rocks for the Pliocene sediments [34]. In contrast, the Pliocene basins south of Mondego river are located to the east, and settle on, Mesocenozoic siliciclastic and carbonate sedimentary units.

The previous-cycle Cretaceous-Cenozoic units from the hinterland basin result mainly from weathering and erosion of Variscan rocks. Their clast framework is composed mainly of variable amounts of quartz, feldspar and fragments of quartzite and granitic and metapelitic rocks [35]. The clay-assemblages comprise kaolinite, mica (or illite), smectite and palygostkyte [35, 36], although kaolinite is the most common mineral in several formations [35, 36, 37]. The geochemistry of these sediments is also highly variable, namely, in regards to the ratio Al_2O_3/K_2O, although CaO and Na_2O were always found in limited proportions (Figure 6). The significant proportions of either potassium feldspar or kaolinite in the Cretaceous-Cenozoic deposits suggest that they integrate mainly sediment derived from the Variscan granitoids. However, given the observed wide compositional range, other source areas may be postulated, in particular in some locations near metapelitic exposures where these rocks may have been significant sediment suppliers. Further works are necessary to better constrain the most likely source units.

Despite the uncertainties regarding the previous-cycle source units, these deposits already contained significant amounts of kaolinite or highly weathered feldspar. Thus, as chemical change will lead to the formation of kaolinite and leaching of most alkaline earth and alkalis elements, under similar climatic conditions it is expected an enrichment of Al_2O_3 south of Mondego.

Physiography could also contribute to justify the very high proportion of illite in the northern sector. These Pliocene deposits are placed west of a fault zone that has been active during the upper Miocene [38] and Pliocene-Pleistocene, [39, 40, 41] and was responsible for the uplift of coastal mountains. This uplifting would eventually be responsible for steep slopes and reduced weathering in the source areas of these deposits. Significant tectonic activity has also been proposed for the sector south of Mondego river [42], but in this region no major structural lineament, comparable to the Porto-Tomar fault zone north of Mondego, is found between the catchment areas with higher relief and the subsiding basins. The Vale Grande samples (whose mineralogy and geochemistry are substantially different from the other northern deposits and are more akin to the southern deposits) are thought to be related to a previous phase when the hinterland drainage was more important [34]. The shift in composition has implications for the timing of the tectonic movements along the Porto-Tomar fault zone and the uplifting of the coastal mountains. If the uplifting of the basin edge was responsible for the differentiation between the older, Al_2O_3 and kaolinite enriched beds, and the younger, K_2O and illite enriched beds, it must be assumed a phase of significant activity in the Porto-Tomar fault zone during the late Pliocene.

Conclusion

The mineralogy and geochemistry (major elements) of the Pliocene units north and south of the Mondego river are substantially different. The sediments of the northern sector are very rich in quartz and illite and have higher amounts of K_2O; sediments of the southern sector are especially rich in kaolinite and have lower amounts of K_2O. These differences suggest that chemical change through exogenous process affected differently the source areas north and south of the Mondego. As these units are considered coeval and deposited under similar climatic conditions, the observed differentiation must relate to the geology of source areas. Physiographic features of the sedimentary environments imposed by tectonic activity may have also played an important role in the differentiation found between the sectors south and north of Mondego river. Important tectonic movements justify a shift in provenance found in the northern sector, when the hinterland drainage was replaced by shorter streams from the basin edge. In addition to this provenance control, local mineralogical and geochemical features should be conditioned by sediment sorting.

Acknowledgments

We are grateful to Laboratório do INETI de S. Mamede Infesta, Porto, Portugal, where the chemical analysis were performed. The paper benefited from valuable discussions with A. F. Soares and P. P. Cunha.

References

[1] H. Nesbitt and G. Young, *Nature* 299, 715 (1982).

[2] H. Chamley, *Clay sedimentology*, Berlin.Springer-Verlag (1989).

[3] A. Singer, *Earth-Science Reviews* 15, 303 (1980).

[4] A. Singer, *Earth-Science Reviews* 21, 251 (1984).

[5] S. M. McLennan, *Reviews in Mineralogy* 21, 169 (1989).

[6] A. C.Morton and C. R. Hallsworth, *Sedimentary Geology* 124, 3 (1999).

[7] B. Bauluz, , M. J. Mayayo, C. Fernandez-Nieto, J. M. Lopez, *Chemical Geology* 168, 135 (2000).

[8] A. Saéz, M. Inglès, L. Cabrera, , A. Heras, *Sedimentary Geology* 159, 305 (2003).

[9] J. M. G. López, B. Bauluz, C. Fernández-Nieto, A.Y. Oliete, *Chemical Geology* 214, 1 (2005).

[10] J. K Tripathi, P. Ghazanfari, V. Rajamani, S. K Tandon, *Quaternary International* 159, 119 (2007).

[11] J. B. Borges, Y.Huh, S. Moon, H. Noh, *Chemical Geology* 254, 52 (2008).

[12] P. Singh, *Chemical Geology* 266, 242 (2009).

[13] M. Thiry, *Earth-Sciences Reviews* 49, 201 (2000).

[14] G. Pe-Piper, L. Dolansky, D. J. W. Piper, *Sedimentary Geology* 178, 75 (2005).

[15] P. A. Dinis, A. F. Soares, *Geologica Acta* 5, 177 (2007).

[16] P. P. Cunha, *Estratigrafia e sedimentologia dos depósitos do Cretácico Superior e Terciário de Portugal Central, a Leste de Coimbra*, PhD thesis, University of Coimbra (1992).

[17] F. T. Rocha, *Argilas aplicadas a estudos litostratigráficos e paleoambientais na Bacia Sedimentar de Aveiro*, PhD thesis, University of Aveiro (1993).

[18] D. Pereira *Sedimentologia e estratigrafia do Cenozóico de Trás-os-Montes oriental (NE Portugal)*, PhD thesis, University of Minho (1997).

[19] A. M. G. Carvalho, *Revista da F.C.U.L* 17, 197 (1972).

[20] B. P. Barbosa, *Est. Notas e Trabalhos do Serv. Fom. Min.* 25, 193 (1983).

[21] P. P. Cunha, B. P. Barbosa, R. P. Reis, *Ciências da Terra* 12, 35 (1993).

[22] J. Grade, C. *Bol. Soc. Geol. Portugal* 22, 197 (1980-81).

[23] V. B. Pereira, *Boletim de Minas* 30 353 (1993).

[24] P. A. Dinis, *Comunicações Geológicas* 93, 81 (2006).

[25] L. G. Schultz, *U.S. Geol. Surv. Prof. Pap.* 391-C, 1 (1964).

[26] D. M. More and R. C. Reynolds, *X-ray Diffraction and the identification and analysis of clay minerals*. Oxford University Press (1997).
[27] M. Kahle, M. Kleber, R. Jahn, *Geoderma* 109, 191 (2002).
[28] J. R. Price and M. A. Velbel, *Chemical Geology* 397 (2003).
[29] L. Harnois, *Sedimentary Geology* 55, 319 (1988).
[30] A. F. Soares, *Estudos do Quaternário* 2, 7 (1999).
[31] E. Garzanti, S. Andò, G. Vezzoli,. *Earth and Planetary Science Letters* 273, 138 (2008).
[32] E. Garzanti, S. Andò, G. Vezzoli, *Earth and Planetary Science Letters* 277, 422 (2009).
[33] H. von Eynatten, R. Tolosana-Delgado, V. Karius, *Sedimentary Geology* (in press, doi: org/10.1016/j.sedgeo.2012.03.008).
[34] P. A. Dinis, A. Oliveira, F. T. Rocha, M. Vieira, P. P. Cunha, *Chemie der Erde* 71, 267 (2011).
[35] P. P. Cunha, *Comunicações Instituto Geológico e Mineiro* 86, 143 (1999).
[36] A. F. Soares, R. P. Reis, S. Daveau, *Memórias e Notícias, Publ. Mus. Lab. Min. Geol. Univ. Coimbra* 96, 3 (1983).
[37] S. Daveau, P. Birot, O. Ribeiro, Les bassins de Lousã et Arganil, *Mem. Centro Estudos Geográficos* 8, 1 (1985-86).
[38] A. J. Sequeira, P. P.Cunha, B. M. Sousa, *Com. Inst. Geol. Mineiro* 83, 95 (1997).
[39] A. B. Ferreira, *Zeitschrift fur Geomorph, v. Suppl. Bd.* 82, 73 (1991).
[40] J. Cabral, *Mem. Instituto Geológico e Mineiro* 31, 1 (1995).
[41] P. A. Dinis, J. M. Cabral, A. F Soares, *Geodinamica Acta* 20, 415 (2007).
[42] A. Ramos, *O Pliocénico e o Plistocénico da Plataforma Litoral entre os paralelos do Cabo Mondego e da Nazaré*, PhD thesis, University of Coimbra (2008).
[43] L. M. F. Simões, M. M. Godinho, L. J. P. F. Neves, *Memórias e Noticias, Pub. Dep. Cienc. Terra e Mus. Min. Geol. Univ. Coimbra* 117, 63 (1994).

In: Advances in Geochemistry Research
Editor: Jorge Sanjurjo-Sánchez ISBN: 978-1-62618-245-5

Chapter 3

GEOCHEMISTRY OF THE BUILT ENVIRONMENT: ALTERATION PRODUCTS OF STONY MATERIALS

Carlos Alves
Centro de Investigação Geológica,
Ordenamento e Valorização de Recursos
(Supported by Fundação para a Ciência e a Tecnologia
with Portuguese Funds, PEst-OE/CTE/UI0697/2011),
Universidade do Minho, Campus de Gualtar, Braga, Portugal

ABSTRACT

It is presented a review addressing the geochemical evolution of materials surfaces in the built environment, focusing namely on the geochemical diversity of decay products. The same geochemical weathering processes that affect natural rock outcrops act on stony materials after their emplacement on the built structures, enhanced by higher concentration of pollutants related to the urban environment and to specific geometric configurations of architectural works. These processes lead to the formation of alteration products that could cause the mobilization of substances of the original materials and the formation of new substances. Two major categories of alteration products will be considered in this review: coatings and erosive features. Coatings can result from neoformations related to pollutants and the reactions between pollutants and the substrate. In the discussion of erosive features will be included processes that imply the loss of material or that create conditions for that loss (fissuring). These erosive features can result from several processes that involve the chemical instability of the original phases of the materials, as well as physical consequences related to the interaction with pollutants, such as expansion and crystallization inside the pore network. The alteration products of stony materials in the built environment present specific geochemical aspects related to the specific conditions and its study is paramount for the definition of qualitative and quantitative relationships between pollutants and alteration features.

INTRODUCTION

The same geochemical weathering processes that affect natural rock outcrops act on stony materials after their emplacement on the built structures, enhanced by higher concentration of pollutants related to the urban environment and to specific geometric configurations related to architectural works. These processes lead to the formation of diverse alteration products. In general lines these alteration products include erosive features and coatings and stains. Detailed description of the alteration products can be easily found in [1-2].

The alteration processes that affect stony materials in the built environment could involve the mobilization of substances of the materials and the formation of new substances that were not present on the materials before emplacement. These alteration products are, therefore, amenable to studies by the usual geochemical tools employed in the natural environment.

The extension of geosciences studies in general to the anthropogenic environment might be considered a polemic subject. For example the International Mineralogical Association has ruled that crystalline substances resulting from the action of geological processes on anthropogenic substances should not be considered as minerals given the diversity and exoticism of anthropogenic new compounds that could be exposed on the natural environment [3]

Nickel [3] also refer that substances formed by the exposure of rock and minerals can be considered as mineral if that exposure was not done with the express purpose of creating new minerals. It could also be considered the question of distinguishing the contribution from natural inorganic (geogenic) sources and anthropogenic sources that is posed in coastal regions and in places with potential volcanic influx [4]. In this case the discussion will concern the identification of what results from human activities or organic processes and cannot be included, hence, as minerals.

Furthermore, one of the major problems that can be found in studies of the built environment concerns the distinction between aspects that result from the exposition of the materials and those that represent deliberate man-interventions, especially in relation to coatings and stains. However, it is already a tradition to use geochemical tools in the research of the effects of human pollutants on the natural environment and this is just an extension to the built environment. The present author contends that the main aspect to be considered is precisely the lack of purpose. Hence, one could include in the geochemical processes the inorganic substances that result from the action of natural processes without human (rational) direction, to wit, without intent. In that sense, the substances whose formation occurs after emplacing in a given construction will be considered here. In relation to the exoticism of the substances it is unclear whether anthropogenic processes can create substances that more exotic than natural processes. One can also considerer that the geochemical tools are apt for the understanding of the complex systems that affect the built environment and that involve several agents acting under variable surrounding conditions. Another problem that could be put on these studies is the distinction of the substances that result from the position of the materials in the built environment from substances that are in the materials. For example, in the same built limestone one could have the original calcite from the sedimentary basin and calcite neoformed in the built environment.

The present chapter attempts to review the geochemical evolution of materials surfaces in the built environment, focusing namely on the geochemical diversity of decay products and the qualitative and quantitative relationships between pollutants and alteration features based on in-field (real built elements) studies. In the next section will be presented an overview of the main geochemical features of alteration products in the built environment, followed by a section discussing the qualitative and quantitative relationships of these alteration features with pollutants.

Geochemical Diversity of Alteration Products

The presentation of the main geochemical features of the alteration products in the built environment considers initially coatings and stain types and then the erosive features in which there is loss of material.

Black Crusts

The expression black crusts in studies on the built environment usually refers to gypsum aggregates that gain the characteristic dark colour from the retention of atmospheric pollution particles as is illustrated in Figure 1 (these black crusts should not to be confounded with Humboldt‘s famous black crusts of the ―equinoctial regions of America" found on natural outcrops and formed by iron and manganese oxides and should also be distinguished from darkish biocoatings). According to what was possible to gather from Google Books, these kind of coatings were reported as early as 1864 [5] as ―the characteristic black crusts which are frequently seen on buildings" made with ―oolitic limestones" and this publication also reported the transformation of calcium lime in calcium sulphate as well as the presence in the crust of ―coal-ashes" (the discussion on the paper presented in the same publication seem to suggest that this was a known problem). While it was an initial trend to consider that black gypsum crusts were specific of calcite-rich rocks (namely as source for calcium) it has been now clearly established that these gypsum crusts can be commonly found in almost any rock type including calcite-poor sandstones [6] and granites [7-10].

Figure 1. Illustration of black crusts affecting large portions of façades.

The black colour results from the fixation of dark atmospheric pollution particles but the formation of gypsum crusts might essentially result from the solubility characteristic of gypsum and be a more widespread process. Kramar et al. [11] report the formation of black gypsum crust on the outside and white gypsum crust inside.

The mineralogy of these crusts is relatively simply with gypsum as dominant phase but with the possible presence of an assortment of other minerals from the substrate or air transported. Some examples include mostly the presence halite [12-16] but also of melanterite (ferrous heptahydrate sulphate), magnesium pentahydrate sulphate [12], mirabilite [17], syngenite [16], the formation of authigenic calcite and dolomite and the reference to peculiar phases such as protodolomite (see [18]).

Geochemically the crusts are dominated by calcium and sulphates but they might present variations that could be related to the involving atmospheric conditions and this would allow its study in a perspective of passive monitoring both in relation to inorganic phases [19] and organic compounds [20].

White Carbonate Crusts

White, carbonate-rich crusts are also frequent on the built environment. This expression has been applied to situations with different distributions that might correspond to two different situations: calcite crusts overlapped or with a space proximity to a carbonate substrate [21-24] and crusts (as illustrated in Figure 2) that show an extensive distribution and patterns that bear a resemblance to free circulating solutions and that have also received other designations, as reviewed by Alves [25], including calcium carbonate stains [26], lime run [27] and calc-sinter [1]. While clearly dominated by calcite, these crusts could have other mineral phases as in Siegesmund et al. [24] that indicated the presence of gypsum.

Siegesmund et al. [24] reports that white crust (in this case associated with carbonate rocks substrates) are the most common weathering crusts in Budapest (Hungary).

According to the experience of the present author in Portugal, while darkish gypsum crusts are frequently found even in places without a high industrial pollution, white carbonate crusts are also extremely common and might cover much wider areas and present a wider distribution in relation to architectural features [see 28-29].

Figure 2. Illustration of carbonate-rich whitish coatings.

They might, furthermore, present also the worrisome (from a cleaning point of view) feature of recurrence (see [25]). Alves and Sanjurjo [30] have speculated on the possible environmental monitoring potential of this type of coating in analogy with what has been done with natural carbonate deposits [31]. In the context of architectural works, namely moderns ones, the presence of white crusts inside the constructions will be useful as marker of infiltration [29] and will help to access defects of the constructions under what can be considered normal conditions of use [32]. The expression ―white crusts" could also apply to crust-like aggregates of other soluble salts (see [33]) that have, however, a more limited space extension and that will be included in the discussion below in relation to soluble salt efflorescences.

Other Coatings and Stains

While being possibly the most common alteration feature, moisture stains will be less interesting from the geochemical point of view since it mainly represents the increase in water. However it deserves some attention given that water is an important decay agent per se, a promoter, a carrier of other pollutants (the specifications of water migration pathways affects the geochemistry of the generality of alteration processes) and also a promoter of alteration processes.

In this section will be also considered salt efflorescences that result from the precipitation of salts from solutions on the surface of the materials and that can lead to whitish stains such as those illustrated in Figure 3. There are several phases that can be found in salt efflorescences at the built environment and a summary can be found in Arnold and Zehnder [33]. They correspond to phases that can usually be found in evaporite deposits but there are some differences that will be highlighted.

A striking difference concerns the importance of nitrate salts in efflorescences that in the natural environment are almost confined, as important phases, to the Chilean salars but that are widespread in the built environment, namely in older constructions, being also be found in some modern constructions (see [34]) and. Even in relation to the nitrates a difference could be highlighted: the predominance of niter (KNO_3) in relation to nitratine ($NaNO_3$) especially in older construction, reflecting the difference in the pollution sources characteristics and the higroscopicity of these salts.

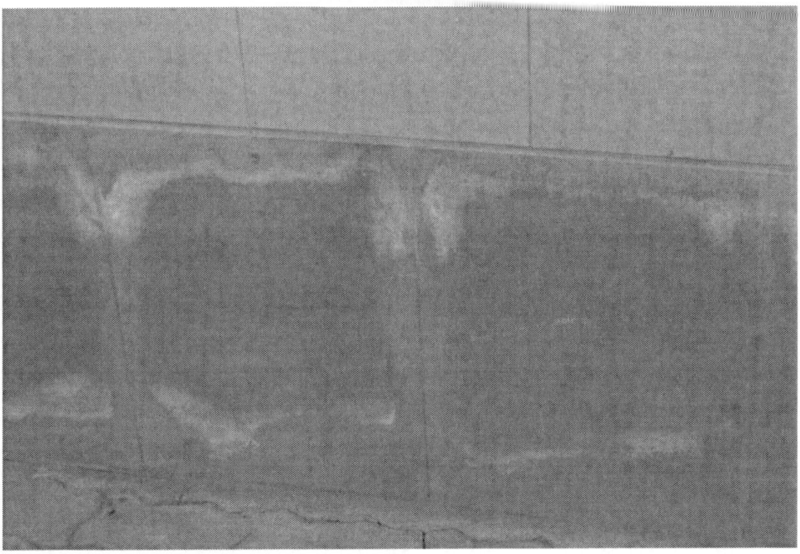

Figure 3. Example of salt efflorescences causing whitish stains in a modern construction.

Another potential difference concerns the absence of certain phases in the built environment. While some peculiar salts that are found in the Atacama deposits of Chile such as darapskite have been also found in monuments [35-36], the present review did not found references to occurrences in the built environment of certain soluble salts that are found in evaporate deposits with a certain frequency, such as iodates (mostly confined to the Chilean nitrate deposits) and borates. One could speculate that this could be seen as supporting the hypothesis of non-marine and inorganic sources for the occurrence of these salts in natural evaporates hence suggesting a possibly further point of interest for the geochemistry of salt systems in the built environment.

There can be found patterns of distribution of soluble salts at the level of the individual building element and at the level of the whole monument and several detailed examples can be found at Alves and Sequeira Braga [36]. Studies regarding a more global perspective on distribution of the mineral substances are scarcer with some examples regarding towns or monuments on a region being found in Alves et al. [37] and Antunes et al. [38]. These studies show the contrast in efflorescences composition between historical centres of continental towns of Portugal (dominated by alkaline nitrates and chlorides) and the Azores Islands (dominated by sodium sulphate and carbonate and where nitrates and magnesium phases were not detected). There is a study by Nord [39] that makes a synthesis of results in a country (Sweden) concluding that sulphates like gypsum and thenardite were the most frequent salts with chlorides and nitrates being less usual.

As in any other study of issues sensible to environmental conditions, the sampling conditions, specifically weather conditions, need to be considered (see [40]) and the usual conditions on the built environment could hinder the identification of more hygroscopic evaporitic phases.

Another type of neformation substances that have received an enormous amount of attention in the built environment are oxalates that can occur in different stone types and in different settings [41-43].

Other neoformations with a significative frequency are phosphates [44-48]. There are also some references to silica-rich coatings [13, 27, 49-50] and coatings enriched in iron and other metals [13, 51-54].

There also some references to other peculiar phases corresponding to specific situations, like formate salts [55] and a high diversity of neoformation phases related to the alteration of metallic elements that will not be discussed here given its particular context (examples of neoformations related to alteration of metallic elements can be found in [56]).

Erosive Features

Alteration processes could create conditions for the loss of material from the surface and back reliefs (as shown in Figure 4) and in this section will be discussed the neoformation phases that occur associated with these erosive products (and therefore might contribute to loss of material).

The main geochemical agents associated with erosive features are soluble salts. Gypsum deserves special attention since it has been found associated with erosive processes in diverse situations, either with black crusts or in features such as granular disintegration or planar detachments (some examples include [7, 12, 14, 36, 45, 57-62]).

In the case of planar detachments there seems to be a preference for the concentration of gypsum at the detachment interface from the substrate. The predominance of calcium and sulphate is also frequently observed in erosive features [7, 36, 57-58, 63]. Examples of other soluble salts that have been associated with erosive features include halite [7, 12, 36, 45, 58, 61, 64-66], magnesium sulphate [67-70], alkaline sulphates and carbonates [39, 62, 71], niter and nitratite [45].

Some more exotic substances can be also found such as nickel sulphate [60] that the authors associated with the characteristics of the stone. Chemical analyses of erosive features show dispersion of compositions (some examples could be seen in [63, 72-73]. There is also a frequent erosive decay process of mortars and concrete materials related to the presence of sulphates such as gypsum, ettringite and thaumasite designated as sulphate attack [74-79], and that has also been considered for bricks [78].

A particularly complex problem is the possible presence in the alteration products of neoformation silicates, reflecting chemical weathering of the stones after their emplacement in the built environment (with the additional interest of being a possible source of substances for coatings and stains – see next section).

However, since in general these silicates could be formed in the natural environment there is the problem of assessing whether the detected phases are indeed neoformations on the built environment or a feature "inherited" form the quarry, a problem that is exacerbate by the usual limitations in sampling that affect the studies of the built environment.

Examples of authors that have proposed the neoformation of silicates in the built environment include Schiavon [80] and García-Guinea et al. [62].

Figure 4. Illustration of the effect of erosion on stone blocks with marked backrelief. Note also the erosive pattern in height, typical of capillary-rising solutions (as would be discussed later on in the text).

QUALITATIVE AND QUANTITATIVE RELATIONS WITH POLLUTANTS

A first point to consider is the possible unequivocal association of a given neoformation phase with a given pollution source, a matter that is relevant for understanding the alteration processes and also for the planning of possible conservation procedures. This is problematic in situations where it is admitted the influence of several sources bearing the polluting substances related to the neoformations phases (as can be seen in the polemics of the Chilean salars) and relates to the more general epistemological problem of equifinality (see Turkington and Paradise [6] and references therein).

It is the kind of situation that arises frequently in the studies of the built environment where besides the overlapping effects of naturally-driven processes there also contributions from past and present human procedures (see examples in Arnold, [81]). Some of the most common sources and pathways of pollutants in the built environment are discussed in Arnold and Zehnder [33] and in Sanjurjo and Alves [82] and might include, groundwater, soilwater and dry and wet deposition and also the materials themselves (as was already indicated by Vitruvius that referred that mortars prepared with sand beach could cause the formation of efflorescences). Another feature to consider is the role played by micro-organisms (see a review of the contribution of biological agents to salt efflorescences in Laiz et al. [83]).

One can begin with the case of water since, besides being a pollutant itself, it is the main carrier of pollutants. In general the observation of the patterns of distribution of decay features could give an important contribution to the distinction between infiltrations, capillary-rising and other modes of water penetration points and pathways (see for example [29, 82].

In some cases, like the formate salts studied by Zehnder and Arnold [55], their extreme rarity allows its association with the contribution of specific pollution sources such as previous cleaning processes. Another example of rarer phases that are attributed to specific characteristics (in this case to a relative rarer type of rock) is the case of nickel sulphate [60]. But in other cases there must be care with the apparent obvious source. While perhaps most people will think of marine contributions in the presence of halite, there could be important sources from human activities (as expressed by Arnold [81], ―Chlorides near the sea may not be from the sea"). Ettringite and thaumasite are usually considered indication of sulphate attack but the source of the sulphates might be diverse [74-79]. Niter (KNO_3) is usually taken to be associated with biogenic contributions [33, 81], however there are authors that admit its origin from the weathering of K-bearing phyllosilicates [11]. The presence of alkaline sulphates and carbonates is usually attributed to the use of modern alkaline cements [33] but there are also authors that consider contributions from the weathering of alkali feldspar [62]. Magnesium sulphates are frequently attributed to alteration of dolomitic components (33, 38, 67) but the contribution from hydrolysis of silicates has also been proposed the [67].

One of the major controversies in this field correspond to the presence of oxalates and the distinction between the result of the action of organisms from those occurrences that are attribute to vestiges of past treatments [41, 43, 61]. A similar controversy have arisen in relation to phosphate neoformations that have been attributed to organic contributions [44-45] and to atmospheric pollution [46] but also studies to past treatments [47].

Iordanidis et al. [48] refers the presence of phosphates in paintings and considered the hypothesis of migration from plasters with organic matter. It is interesting to note the presence of phosphate occurrences in indoor recent constructions as is illustrated in Figure 5.

In relation to the whitish carbonate-rich crusts it seems possible to consider two main situations that correspond to the distribution patterns presented above.

There are crusts that are attributed to the weathering of a carbonate substrate and the reprecipitation of calcite [21-24]. Siegesmund et al. [24] further indicated the presence of gypsum on these white calcite-dominated crusts. There are also references that attribute the presence of calcite crusts to biogenic weathering [84].

a

Figure 5. (Continued).

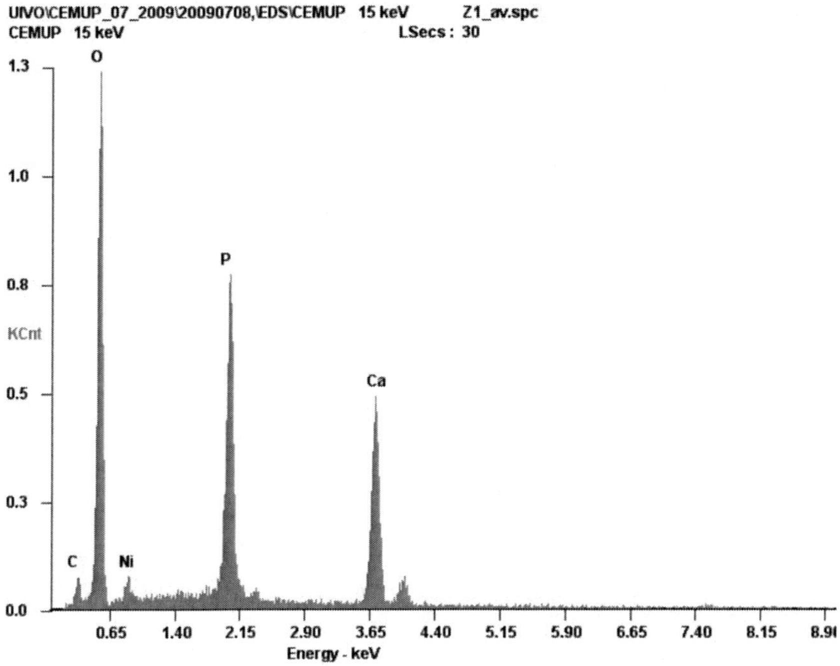

b

Figure 5. Scanning electron microscopy studies of phosphates on a modern construction (a- image; b – spectrum). Studies developed at the CEMUP laboratory of the University of Oporto, Portugal.

There are also crusts related to flowing and infiltrating solutions (see Figure 2) apparently related to solutions from the mortars that sometimes can present speleothem-like and tufa like features and that can receive several designations, as reviewed by Alves [25]. There are some questions remaining in this subject since while carbonate crusts could be related to the initial solutions of the cements, their thickness and recurrence (as the present author has observed in some places) seem to indicate that there are contributions from present sources (presumably dissolution of mortars phases), an aspect that can have relevance in terms of cleaning.

The ubiquitous occurrence of gypsum has generated an extensive debate on its genesis and, besides the classical ―calcium from carbonate rocks stones" and ―sulphur from atmospheric pollution" a diversity of sources has been proposed for the calcium and the sulphur, including marine influence [12], volcanic emissions [4], leaching from mortars [7, 58, 87], sulphates already present in the rocks [33, 85-86], the weathering of feldspars [8] and actinolite

[88], weathering of sulphides [49, 89-92], weathering of previous coatings with gypsum [10] and biological activity [42, 93]. This multitude of possible sources has promoted intensive research using a panoply of geochemical tools (see review in Sanjurjo and Alves [82]).

Besides the direct formation from a given source it is also necessary to consider differentiation processes in a similar way to what has been used in studies of natural evaporites. The chemical divides proposed by Hardie and Eugster [94] has been considered for evolution of salt solutions in the built environment with the incorporation of nitrate (an insignificant ion in seawater) by Steiger [68]. However, in the built environment there are frequent situations where the initial composition is biased towards certain ions due to specific pollution sources. In the general case (as is shown by Steiger [68]) gypsum is a salt that tends to precipitate easily, but that is also sufficiently soluble to be mobilized to a certain distance and therefore spread erosive decay (see Sanjurjo et al. [95]). Differentiation processes have been invoked (in analogy to what is observed in seawater evaporation) to explain the occurrence of magnesium sulphates [36] and the formation of dolomite and protodolomite associated with gypsum precipitation [18]. The geochemical complexity that affects the behaviour of the salt processes is also strongly related to the environmental conditions (given the higroscopicity of most soluble salts) and a thermodynamic model [95] has been proposed for prediction the behaviour of salt mixtures (Runsalt (c) - http://science.sdf-eu.org/runsalt/).

Similarly to what was referred for water, the study of the distribution of the soluble salts occurrences could be an important indicator in the discussion of the sources (see, e.g., [36, 38, 63, 97]) namely by comparison of situations (in a perspective akin to ―differential diagnosis"). A well-known example is the typical pattern in height associated with capillary-rising solutions (see Figure 4). It is also necessary to collect the available historical information since it could shed light on possible presence of specific pollution situations. In cases where it is possible to establish associations between pollution sources and neoformation phases, there is the additional possible interest of considering the neoformations as elements of historical reconstruction (in the sense that these neoformations can be indicators of certain kind of activities). Recent constructions could be another important factor in the study of alteration processes given the lesser overlap of pollution events [30].

The beginning of erosive processes related to the action of pollutants such as soluble salts can be relatively fast and Blanco Varela et al. [60] report the occurrence of blisters (with the presence of gypsum and nickel sulphate) that could be observed after one year. The erosive effect of soluble salts have been related to the kind of salt and the associated pressures resulting from crystallisation [64] and hydration [98], to cycles of dissolution-crystallisation resulting from higroscopicity [33] and to solubility and drying conditions [99] that affect the crystallisation position (namely whether salt crystallisation occurs at the surface or inside the pore network related). As has been above, gypsum is frequently related to the occurrence of erosive features and that could be related to its ubiquity and solubility properties [68, 99]. There is the indication that other more soluble salts will tend to present efflorescences instead of crystallisation inside the pore network [11, 36, 99]. But, as was presented above, there are other soluble salts that have been detected associated with erosive features and Nord [39] proposed that soluble salts such as sodium sulphate and sodium carbonate that form crystalline hydrates with a high water content have a large damaging potential. (Sodium sulphate is in fact used in laboratory salt weathering tests of materials due to its high erosive power. Conversely, a case that has deserved some attention is the divergence between the frequent association of gypsum with erosive features and its lower erosive impact (when compared with other salts) in laboratory tests (see Goudie and Viles [100] for an extensive discussion). Of course, as is commonly referred in statistics, association does not mean causality, and the processes of differentiation could have some impact on this matter. Silva et al. [66] referred the increase of gypsum solubility due to the presence in solution of more soluble salts, a process that favour a sub-superficial crystallisation of gypsum and the development of granular disintegration. Alves and Sequeira Braga [36] proposed the hypothesis of differentiation between erosive features and salt efflorescences on the same stone.

Besides the erosion caused by physical processes related to crystallisation and hydration, pollutants have been associated with disruptive effects due to chemical processes (see reviews in Alves [101] and in Sanjurjo Sánchez and Alves [82]). These processes include alkali-aggregate reactions in concrete (reactions between alkalis and some types of aggregates) and diverse processes of dissolution including carbonates by acid solutions, silicates by acid and basic solution, dolomite by sulphate solutions and by Ca-rich solutions and several other examples related to high ionic strength of salt solution.

Salt solutions can also be associated with fissuring and disruption of stony materials by promoting the deleterious behaviour of clay minerals due to wetting and drying cycles [102] and increasing volume expansions of clays [103]. Salt solutions have additionally been referred as potentially affecting the mechanical strength of the rocks [104-105].

Several coatings rich in iron and other metals referred above have been related to the fixation of atmospheric particles which has also been considered as a possible source for the genesis of silica crusts [13, 50]. However, it is necessary to assess the possible contribution of the weathering of the substrate [49-50, 106]. In the case of the silica crusts, some publications [27] refer that sometimes these crusts are known as ―white scum" and relate them to past cleaning procedures. It will also relevant to remember that silica became increasingly soluble in alkaline solutions [65] such as the pore solutions of cement. An example of silica crusts is presented in Figure 6 and the pattern suggests flowing solutions from the inside of the stone work.

The neoformations could also contribute to erosion by constituting zones with different characteristics that can promote salt crystallization and detachment (examples include [17, 22-23, 49, 65, 107, 108-110].

While there are several proposals of laboratory based to establish damage or dose-response functions in relation to gaseous pollutants (see e.g. [111, 112]) or variations of salt solutions composition (see e.g. [99, 113-114], there are still few indications for validation of these relations based on field evidence.

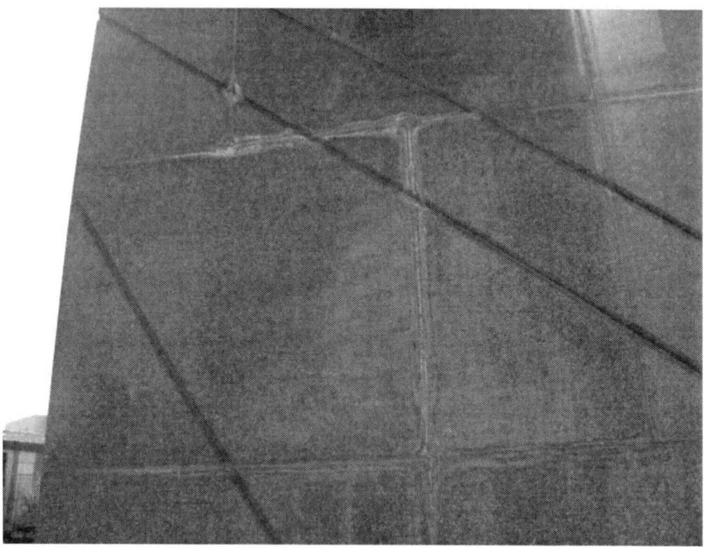

a

Figure 6. (Continued).

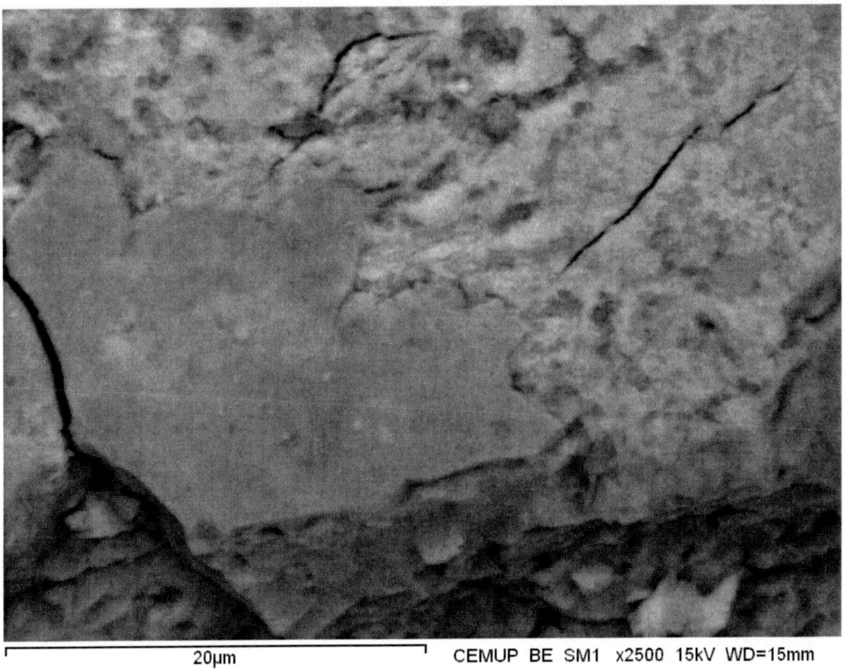

b

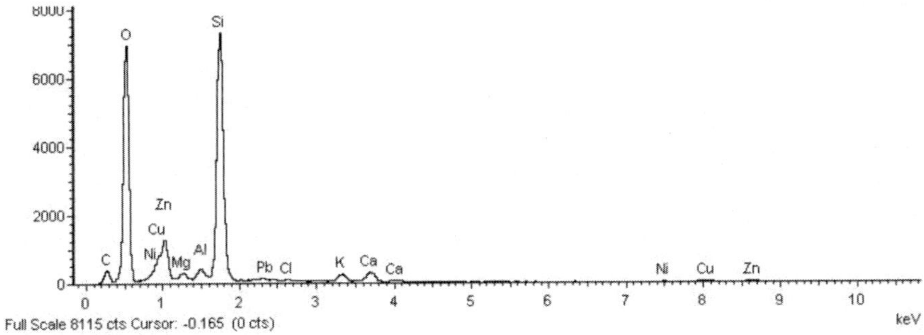

c

Figure 6. Silica coatings on a stonework of the 80's of the last century with field observation (a) and scanning electron microscopy image (b) and chemical spectrum (c). Scanning Electron Microscopy Studies developed at the CEMUP laboratory of the University of Oporto, Portugal.

For example, it has been at least more than a century since the noting of the formation of dark gypsum crusts on monuments and while there are authors like Nord et al. [115] that refer a positive correlation between SO_2

levels and gypsum formation on calcareous stone, it is still unclear what will be the necessary levels of pollutants necessary to their formation.

In a case studied by Alves and Sequeira Braga [36], where there was a roughly good estimation of the pollution source and its position (and old salt storeroom), it was possible to observe a clear pattern of decay intensity related to the position of the stones in relation to the previous kitchen salt storeroom. A correlation between salt load and the extent of damage can be seen in a study presented by Fitzner and Heinrichs [1]. According to Doehne and Pinchin [69] the dominant factors in damage at the studied site were a combination of environmental condition and local salt concentration. Inhabitants of coastal regions are well aware of the effects of seaspray on metallic materials and there are also indications (including Herodotus and Vitruvius) of higher salt weathering hazards for stony materials in regions near the sea. Mottershead et al. [116] compared two historical structures (one near the coast and another farther inland) concluding that decay rates were higher for the stone in the coastal structure. Stefanis [117] defends that stone decay related to atmospheric deposition will be higher for higher amounts of salts. Costa et al. [118] found that the zone of average aggressiveness for built structures due to seaspray was found around a hundredth and a half meters from the coast line. The present author has observed in mainland Portugal and especially in the Azores Islands distributions patterns of weathering products in the built environment that suggest a marked influence of the proximity to the seaside [37].

The conditions of action of the pollutants can also play a critical part in the decay processes as exemplified in Alves and Sequeira Braga [36] where erosive processes of paintings were related to recursive crystallisation of niter resulting from infiltrations associated with rain events. There are also studies showing a marked contrasts between outdoor and indoor zones due to different exposition to the main decay agents as observed in the Azores Islands (see [37, 73]). Another illustrative example can be found in a study of a monument of mainland Portugal [96] where efflorescences of alkaline sulphates seemed to cause little damage to the granite surface. This result was attributed to the milder environmental conditions that (following to the model proposed in Hammecker [98]) favoured slower drying and surface crystallisation.

However one should not think that indoor zones are free from the salt weathering hazard as this has been observed in stones, especially due to the occurrences of gypsum but also in relation to anomalous amounts of sodium chloride [36]. Indoor erosion of wall paints has been also observed in association with diverse soluble salts [33].

Considering the diverse factors that weight on the relation between pollutants loads and alteration processes, Sanjurjo and Alves [82] suggest that qualitative modelling based on the characterisation of pollutants, design features of built elements and field assessments of previous performances would be the best way to make predictions regarding the alteration of materials on the built environment (for general considerations in relation to qualitative modelling on Earth Sciences see Pilkey and Pilkey-Jarvis [119]).

FINAL CONSIDERATIONS

In this review was presented an outlook of the main geochemical characteristics of the alteration features of stony materials on the built environment, highlighting the diversity of neoformation substances that were present on coatings and the association between erosive features and the presence of soluble salts. There are geochemical trends related to the specificities of the built environment among supergenic environments: the abundance of gypsum and nitrates; the apparent absence of certain evaporite minerals, namely borates. It was speculated that this could also be relevant for discussions on the sources of these substances on other environments.

Regarding the qualitative and quantitative relations between the pollutants and alteration features, it was seen that while gypsum is the soluble salt most commonly associated with erosive features, several other salts have been also identified in these weathering products. It was also highlighted the problems that affected the establishment of the sources of the pollutants given that generally there are several potential sources were the pollutant can be found. Besides the potential overlap of diverse pollution sources one needs also to consider differentiation processes of the pollutants that occur in the built environment. Besides diverse advanced analytical techniques, the understanding of the pollution processes can be generally aided by the use of the general geochemistry epistemic approach, to wit, the consideration of the analytical chemical results in a spatial (geo) perspective, namely the distribution patterns of the pollutants.

The relations between pollutants loads and development of alteration features were also considered. In relation to the behaviour of materials in real structures our present knowledge is generally limited to rough relations of the kind ―high load implies intense decay". This is still a poorly developed field and a most relevant one in conservation terms, namely regarding the control of pollution emissions and the levels that need to be attained in intervention procedures. Again, a global field perspective of existing built structures (including morphological and environmental characteristics) would be essential for a useful understanding of these relations.

REFERENCES

[1] Fitzner, B., K. Heinrichs (2002) Damage diagnosis on stone monuments – weathering forms, damage categories and damage indices. In: *"Understanding and managing stone decay"*, The Karolinum Press, Prague, p. 11-56. Freely available online (May 2012) at http://www.stone.rwth-aachen.de/decay_diagnosis.pdf

[2] ICOMOS-ISCS-International Council on Monuments and Sites-International Scientific Committee for Stone (2008) Illustrated glossary on stone deterioration patterns. Freely available online (May 2012) at http://www.international.icomos.org/publications/monuments_and_sites/15/pdf/Monuments_and_Sites_15_ISCS_Glossary_Stone.pdf

[3] Nickel, E. H. (1995) The Definition of a Mineral. *The Canadian Mineralogist* 33:689-690.

[4] Montana, G., Randazzo, L., Mazzoleni, P. (2012) Natural and anthropogenic sources of total suspended particulate and their contribution to the formation of black crusts on building stone materials of Catania (Sicily). *Environmental Earth Sciences* (online) 1-14. Doi:10.1007/s12665-012-1554-x

[5] Voelcker, A. (1864) On the injurious effects of smoke on certain building stones and on vegetation. *Journal of the Society of Arts* 12:146-153. Freely available (May 2012) at http://books.google.com/

[6] Turkington, A. V., Paradise, T. R. (2005) Sandstone weathering: a century of research and innovation. *Geomorphology* 67:229-253. Doi:10.1016/j.geomorph.2004.09.028.)

[7] Casal Porto, M. (1989) Estudio de la alteración del granito en los edificios de interés histórico de la provincia de La Coruña. *Tesis doctoral,* Univ. Santiago de Compostela. In Spanish.

[8] Alonso Zarza, A. M., García del Cura, M. A., López de Azcona, M. C., Mingarro Martín, F. (1991) Estudio previo de las «costras negras» desarrolladas sobre rocas graníticas en la Catedral de Toledo. *Geogaceta*, 9, 79-81. In Spanish. Freely available online (May 2012) at http://www. sociedadgeologica.es/archivos/geogacetas/Geo09/Art25.pdf

[9] Sequeira Braga, M. A., Simões Alves, C. A., Begonha, A. (1993) Weathering of the Oporto granite and the deterioration of the Hospital de Santo António: Historical monument built with granitic materials. *Alteración de granitos y rocas afines*, C.S.I.C, Spain, pp. 153-154.

[10] Sanjurjo Sánchez, J., Vidal Romani, J. R., Alves, C. A. S., Fernández Mosquera, D. (2009) Origin of gypsum-rich coatings on Historic Buildings. *Water Air Soil Poll.* 204:53-68. Doi:10.1007/s11270-009-0025-9

[11] Kramar, S., Urosevic, M., Pristacz, H., Mirtič, B. (2010), Assessment of limestone deterioration due to salt formation by micro-Raman spectroscopy: application to architectural heritage. *J. Raman Spectrosc.*, 41:1441–1448. Doi:10.1002/jrs.2700

[12] Zezza, F., Macrì, F. (1995) Marine aerosol and stone decay. *Sci. Total Environ.* 167:123-143. Doi:10.1016/0048-9697(95)04575-L

[13] Moropoulou, A., Bisbikou, K., Torfs, K., Vangrieken, R., Zezza, F., Macri, F. (1998) Origin and growth of weathering crusts on ancient marbles in industrial atmosphere. *Atmosph. Environ.*, 32:967-982. Doi:10.1016/S1352-2310(97)00129-5

[14] Cardell, C., Delalieux, F., Roumpopoulos, K., Moropoulou, A., Auger, F., Van Grieken, R. (2003) Salt-induced decay in calcareous stone monuments and buildings in a marine environment in SW France. *Construction and Building Materials* 17:165-179. Doi:10.1016/S0950-0618(02)00104-6

[15] Sanjurjo Sánchez, J., Vidal Romaní, J. R., Fernández Mosquera, D., Alves, C. A. (2008) Study of origin and composition of coatings in a monument built with granitic rocks, by SEM, XRD, XRF and DTA-TGA. *X-Ray Spectrometry* 37:346-354. Doi:10.1002/xrs.1019

[16] Mosyagin, A., Knauf, I. V., Zelenskaya, M. S. (2009) Deterioration of carbonate rocks used for archeological monuments in Tauric Chersonesos (Crimea). Studia UBB Geologia 54:13-16. Doi:10.5038/1937-8602.54.2.3. Freely available online (May 2012) at http://scholarcommons.usf.edu/geologia/vol54/iss2/art3/

[17] Smith, B. J., Whalley, W. B., Magee, R. (1991) Building stone decay in a "clean" environment: western Northern Ireland. In: *"Science, Technology and European Cultural Heritage"*, Butterworth Heinemann: Oxford, UK, pp. 434-438.

[18] Rodríguez-Navarro, C., Sebastian, E., Rodríguez-Gallego, M. (1997a) An urban model for dolomite precipitation: authigenic dolomite on weathered building stones. *Sedimentary Geology* 109:1-11. Doi:10.1016/S0037-0738(96)00041-3.

[19] Del Monte, M., Ausset, P., Forti, P., Lèfevre, R. A., Tolomelli, M. (2001) Air pollution records on selenite in the urban environment. *Atmospheric Environment* 35:3885-3896. Doi: 10.1016/S1352-2310(01)00143-1.

[20] Sabbioni, C., Ghedini, N., Bonazza, A. (2003) Organic anions in damage layers on monuments and buildings. *Atmos. Environ.* 37:1261-1269. Doi:10.1016/S1352-2310(02)01025-7

[21] Maravelaki-Kalaitzaki, P., Biscontin, G. (1999) Origin, characteristics and morphology of weathering crusts on Istria stone in Venice. *Atmospheric Environment* 33:1699-1709. Doi:10.1016/S1352-2310(98)00263-5

[22] Fitzner, B., Heinrichs, K., Volker, M. (1996) Model for salt weathering at Maltese Globigerina Limestones. In: *"Origin, Mechanisms and Effects of Salt on Degradation of Monuments in Marine and Continental Environments"*, Protection and Conservation of the European Cultural Heritage Research Report, 4, 331–344.

[23] Cassar, J. A. (2002) Deterioration of the Globigerina Limestone of the Maltese Islands In Natural Stone, Weathering Phenomena, Conservation Strategies and Case Studies, Special Publications 205, Geological Society: London, UK, pp. 33-49.

[24] Siegesmund, S., Török, A., Hüpers, A., Müller, C., Klemm, W. (2007) Mineralogical, geochemical and microfabric evidences of gypsum crusts: a case study from Budapest. *Environmental Geology* 52:385-397. Doi:10.1007/s00254-006-0588-3

[25] Alves, C. (2010) "White" crusts on recent buildings. *Materials Science Forum* 636 – 637:1300 1305. Doi:10.4028/www.scientific.net/MSF.636-637.1300

[26] Beall, C. (2003) Masonry Design and Detailing: For Architects and Contractors. McGraw-Hill Professional, United States of America.

[27] BIA-Brick Industry Association (2006) Stains - Identification and Prevention. TECHNICAL NOTES on Brick Construction, v. 23. Freely available online (May 2012) in http://www.ambrico.com/wp-content/uploads/2010/11/TN23.pdf

[28] Alves, C. (2009c) Widespread occurrence of white coatings in recent buildings. IRF'2009, *3rd International Conference on Integrity,*

Reliability and Failure, Porto, CD-ROM. Freely available online (May 2012) at http://paginas.fe.up.pt/clme/IRF2009/proceedings/papers/P0327.pdf

[29] Alves, C., Figueiredo, C., Maurício, A., Figueiredo, P., Aires-Barros, L. (2009) Contribution of exogenous substances to alteration of stone surfaces in metropolitan stations. IRF'2009, *3rd International Conference on Integrity, Reliability and Failure,* Porto, CD-ROM. Freely available online (May 2012) at http://paginas.fe.up.pt/clme/IRF2009/proceedings/papers/P0316.pdf

[30] Alves, C., Sanjurjo, J. (2011) Geoscience of the Built Environment: Pollutants and Materials Surfaces. *Geosciences* 1:26-43. Doi:10.3390/geosciences1010026

[31] Arenas, C., Osacar, C., Sancho, C., Vazquez-Urbez, M., Auque, L., Pardo, G. (2010) Seasonal record from recent fluvial tufa deposits (Monasterio de Piedra, NE Spain): sedimentological and stable isotope data. In: ―Tufas and Speleothems: Unravelling the Microbial and Physical Controls" *Geological Society Special Publications* 336; Geological Society, London, UK, pp. 119-142.

[32] Alves, C., Figueiredo, C., Maurício, A., Figueiredo, P., Aires-Barros, L. (2011) Performance of stones under different conditions: a study of metro stations. *MATERIAIS* 2011, p. 359.

[33] Arnold, A., Zehnder, K. (1991) Monitoring Wall Paintings Affected by Soluble Salts. *The Conservation of wall paintings*, Getty Conservation Institute, pp 103-135. Freely vailable online (May 2012) at http://getty.edu/conservation/publications/pdf_publications/wall_paintings.pdf.

[34] Alves, C. (2009a) Old salts in modern constructions. IRF'2009, *3rd International Conference on Integrity, Reliability and Failure,* Porto, CD-ROM. Freely available online (May 2012) at http://paginas.fe.up.pt/clme/IRF2009/proceedings/papers/P0474.pdf

[35] Holtkamp, M. H. P. C., Heijnen, W. M. M. (1991) The Mineral Darapskite in the Efflorescence on Two Dutch Churches. *Studies in Conservation* 36:175-178.

[36] Alves, C. A. S., Sequeira Braga, M. A. (2000) Decay effects associated with soluble salts on granite buildings of Braga (NW Portugal) In: *"Environmental Mineralogy: Microbial Interactions, Anthropogenic Influences, Contaminated Land and Waste Management"*. Mineralogical Society of Great Britain and Ireland, London, pp 181–199.

[37] Alves, C. A. S., Cruz, J. V., França, Z. T. T. M. (2003) Estudo de eflorescências salinas em Monumentos e edifícios de São Miguel

(Açores). *Proceedings IV Congresso Ibérico de Geoquímica*, Coimbra, Portugal, pp. 294-296. In Portuguese.

[38] Antunes, S. A. G., Moreno, F., Alves, C. A. S., Pamplona, J. M. V. (2004) Erosive de-cay features of granitic stones monuments and characterisation of salt ef-florescences in Évora (SE Portugal). *Proceedings 6th International Symposium The Conservation of Monuments in the Mediterranean Basin*, Lisbon, pp. 223 –228.

[39] Nord, A. G. (1992) Efflorescence salts on weathered building stone in Sweden, *Geologiska Föreningen i Stockholm Förhandlingar* 114:423-429, DOI:10.1080/11035899209453901

[40] Sawdy, A., Price, C. (2005) Salt damage at Cleeve Abbey, England. Part II: seasonal variability of salt distribution and implications for sampling strategies. *Journal of Cultural Heritage* 6:361-367. Doi:10.1016/j.culher.2005.10.002

[41] Del Monte, M., Sabbioni, C., Zappia, G. (1987) The origin of calcium oxalates on historical buildings, monuments and natural outcrops. *Science of The Total Environment* 67:17-39. Doi:10.1016/0048-9697(87)90063-5.

[42] Blázquez, F., García-Vallès, M., Krumbein, W., Sterflinger, K., Vendrell-Saz, M. (1997). Microstromatolitic deposits on granitic monuments: development and decay. *European Journal of Mineralogy* 9:889-901.

[43] Pavía, S., Caro, S. (2006) Origin of films on Monumental Stone. *Stud. Conserv.* 51:177-188.

[44] Hosono, T., Uchida, E., Suda, C., Ueno, A., Nakagawa, T. (2006) Salt weathering of sandstone at the Angkor monuments, Cambodia: Identification of the origins of salts using sulfur and strontium isotopes. *J. Archaeol. Sci.* 33:1541-1551. Doi:10.1016/j.jas.2006.01.018

[45] Begonha, A. (2009) Mineralogical study of the deterioration of granite stones of two Portuguese churches and characterization of the salt solutions in the porous network by the presence of diatoms. *Materials Characterization* 60:621-635. Doi:10.1016/j.matchar.2008.12.019

[46] Ostrooumov, M. (2009) A Raman, IR and XRD analysis of the deterioration on historical monuments: Case study from Mexico. *Spectrochimica Acta Part A: Molecular and Biomolecular Spectroscopy* 73:498-504. Doi:10.1016/j.saa.2008.12.033.

[47] Polikreti, K., Maniatis, Y. (2003) Micromorphology, composition and origin of the orange patina on the marble surfaces of Propylaea

(Acropolis, Athens). *Science of The Total Environment* 308:111-119. Doi:10.1016/S0048-9697(02)00613-7

[48] Iordanidis, A., Garcia-Guinea, J., Strati, A., Gkimourtzina, A., Papoulidou, A. (2011) Byzantine wall paintings from Kastoria, northern Greece: Spectroscopic study of pigments and efflorescing salts. *Spectrochimica Acta Part A: Molecular and Biomolecular Spectroscopy* 78: 874-887. Doi:10.1016/j.saa.2010.12.055.

[49] Nakhla, S. M., Mahgoub, G. A., Hubacek, H. (2003) Study of the Mechanism of Flake Formation in Limestone: A Step Towards the Stabilization of Stone Structures. *Eighth international congress of Egyptologists* 2000, V. 3, pp. 353-359.

[50] Sanjurjo-Sánchez, J., Vidal Romaní, J. R., Alves, C. (2012) Comparative analysis of coatings on granitic substrates from urban and natural settings (NW Spain). *Geomorphology* 138:231-242. Doi:10.1016/j. geomorph.2011.09.008.

[51] Nord, A. G., Ericsson, T. (1993) Chemical analysis of thin black layers on building stone. *Studies in conservation* 38:25-35.

[52] Begonha, A., Sequeira Braga, M. A. (1996) Black crusts and thin black layers in granitic monuments: their characterization and the role of air pollution. *8th International Congress on Deterioration of Conservation of Stone*. Möller Druck und Verlag, Berlin, pp 371-375.

[53] Schiavon, N. (1996) Soiling of urban granite 1: Microfabrics and mineralogical aspects. In: *"Degradation and Conservation of Granitic Rocks in Monument"*, Protection and Conservation of the European Cultural Heritage Research Report No. 5, European Commission: Brussels, Belgium, pp. 307-312.

[54] Matias, J. M., Alves, C. A. S. (2002) The influence of petrographic, architectural and environmental factors in decay patterns and durability of granite stones in Braga monuments (NW Portugal). In: *"Natural Stone Weathering Phenomena, Conservation Strategies and Case Studies"*. The Geo-logical Society, London, UK, pp. 273 –281.

[55] Zehnder, K., Arnold, A. (1984) Stone damage due to formate salts. *Studies in Conservation* 29:32-34.

[56] Roberge, P. R. (2000) *Handbook of Corrosion Engineering*. McGraw-Hill, New York, US.

[57] Wendler, E., Klemm, D. D., Snethlage, R. (1990) Contour scaling on building facades - Dependence on stone type and environmental conditions. *Advanced Workshop Analy. Method. for the Invest. of Damaged Stone*, Pavia.

[58] Casal Porto, M., Silva Hermo, B., Delgado Rodrigues, J. (1991) Agents and forms of weathering in granitic rocks used in monuments. In: *"Science, Technology and European Cultural Heritage"*, Butterworth Heinemann:Oxford, UK, pp. 439-442.
[59] Jeannette, D. (1992) Morphologie et nomenclature des alterations. In: *"La conservation de la pierre monumentale en France"*, CNRS, France, pp. 51-72. In French.
[60] Blanco Varela, M. T., Menendez, E., Hoyos, M. (1992) Study of surface decay of the marbles and serpentine from the Descalzas Reales Convent at Madrid. *Proceedings 2nd international symposium The conservation of monuments in the Mediterranean Basin*, pp. 167-175.
[61] Lazzarini, L., Borrelli, E., Bouabdelli, M., Antonelli, F. (2007) Insight into the conservation problems of the stone building "Bab Agnaou", a XII cent. monumental gate in Marrakech (Morocco). *Journal of Cultural Heritage* 8:315-322. Doi:10.1016/j.culher.2007.02.002
[62] García-Guinea, J., Almendros, G., Benavente, D., Correcher, V., Pérez-García, A., Recio-Vázquez, L., Sánchez-Moral, S. (2012) Incompatible building materials within the stereotomic Avalos sculptures of the Valley of Fallen (Madrid, Spain). *Materiales de Construcción* [online]. Doi: 10.3989/mc.2012.07011
[63] Moreno, F., Vilela, S. A. G., Antunes, A. S. G., Alves, C. A. S. (2006) Capillary-rising salt pollution and granitic stone erosive decay in the parish church of Torre de Moncorvo (NE Portugal)-implications for conservation strategy. *Journal of Cultural Heritage* 7:56-66. Doi:10. 1016/j.culher.2005.10.006
[64] Winkler, E. M., Singer, P. C. (1972) Crystallization Pressure of Salts in Stone and Concrete. *Geol. Soc. Amer. Bull.* 83:3509-3514.
[65] Winkler, E. M. (1994) Stone in architecture. *Properties, Durability.* Springer-Verlag.
[66] Silva, B., Rivas, T., Prieto, B. (1996) Relation between type of soluble salts and decay forms in granitic coastal churches in Galicia (NW Spain). Origin, Mechanisms, and Effects of Salts on Degradation of Monuments in Marine and Continental Environments. *Protection and Conservation of the European Cultural Heritage Research Report No*. 4, pp. 183-190.
[67] Blanco, J. (1989) Dinámica del proceso de alteración ambiental de la "Piedra de Villamayor. *Geogaceta* 6:32– 35.

[68] Steiger, M. (2003) Salts and Crusts. In: *"The Effects of air pollution on the built environment"*, Imperial College Press, London, UK, pp 133-181.

[69] Doehne, E., Pinchin, S. (2008) Time-lapse macro-imaging in the field: monitoring rapid flaking of magnesian limestone. *11th Int. Cong. on Deterioration and Conservation of Stone* V. 1.

[70] Odgers, D., Pinchin, S., Martin, B., Wood, C., Curteis, T., Doehne, E., Chiari, G., Teutonico, J. M., Burges, A. (2008) Investigations into Decay Mechanisms of Magnesian Limestone at Chapter House, Howden Minster. *11th International Congress on Deterioration and Conservation of Stone,* Vol. 1.

[71] Matović, V., Vasković, N., Erić, S., Srećković-Batoćanin, D. (2010) Interaction between binding materials—the cause of damage to gabbro stone on the monument to the unknown soldier (Serbia). *Environmental Earth Sciences* 60:1153-1164. Doi: 10.1007/s12665-009-0257-4

[72] Torfs, K., Van Grieken, R., Zezza, F., Garcia, N., Macri, F. (1997) The Cathedral of Bari, Italy: Evaluation of Environmental Effects on Stone Decay Phenomena. *Studies in Conservation* 42:193-206.

[73] Alves, C. A. S., Sequeira Braga, M. A., Trancoso, A. (2000) Saline pollution in trachyte monuments of the Azores Islands(Portugal). *9th International Congress on Deterioration and Conservation of Stone,* Elsevier Science B.V., Amsterdam, pp. 225-233. Doi:10.1016/B978-044450517-0/50104-5.

[74] Eglinton, M. (1998) Resistance of concrete to destructive agencies. In: Hewlett, P. C. (ed) *Lea's Chemistry of Cement and Concrete*. Elsevier, Oxford, UK, pp. 299-342.

[75] Santhanam, M., Cohen, M. D., Olek, J. (2001) Sulfate attack research — whither now? *Cement Concrete Res.* 31:845–851. Doi:10.1016/S0008-8846(01)00510-5

[76] Bensted, J. (2002) A discussion of the review paper ―Sulphate attack research—whither now?" by M. Santhanam, M. D. Cohen and J. Olek. *Cement and Concrete Research* 32:995-1000. Doi:10.1016/S0008-8846 (02)00731-7.

[77] Santhanam, M., Cohen, M. D., Olek, J. (2002) Mechanism of sulfate attack: a fresh look. Part 1: summary of experimental results. *Cement Concrete Res.* 32:915-921. Doi:10.1016/S0008-8846(02)00724-X

[78] Larbi, J. A. (2004) Microscopy applied to the diagnosis of the deterioration of brick masonry. *Construction and Building Materials* 18:299-307. Doi:10.1016/j.conbuildmat.2004.02.002

[79] Neville, A. (2004) The confused world of sulfate attack on concrete. *Cement and Concrete Research*, 34, 8, pp 1275-1296. Doi:10.1016/j.cemconres.2004.04.004

[80] Schiavon, N. (2007) Kaolinisation of granite in an urban environment. *Environmental Geology* 52:399–407. Doi:10.1007/s00254-006-0473-0

[81] Arnold, A. (1996) Origin and behaviour of some salts in context of weathering on monuments. Origin, Mechanisms, and Effects of Salts on Degradation of Monuments in Marine and Continental Environments. *Protection and Conservation of the European Cultural Heritage Research Report No. 4,* pp. 134-139.

[82] Sanjurjo Sánchez, J., Alves, C. (2012) Pollutant-Induced Decay of Building Materials. In: *"Environmental Chemistry for a Sustainable World"* Part 1, Springer Netherlands, pp. 47-120, Doi:10.1007/978-94-007-2439-6_2

[83] Laiz, L., Recio, D., Hermosin, B., Saiz-Jimenez, C. (2000) Microbial communities in salt efflorescences. In: *"Of microbes and art: the role of microbial communities in the degradation and protection of cultural heritage"*, Kluwer, New York, US, pp.77-88.

[84] Peraza Zurita, Y., Cultrone, G., Sánchez Castillo, P., Sebastián, E., Bolívar, F. C. (2005) Microalgae associated with deteriorated stonework of the fountain of Bibatauín in Granada, Spain, International Biodeterioration andamp; *Biodegradation* 55:55-61. Doi:10.1016/j.ibiod.2004.05.006

[85] Heiss, K., Skalli, M., Zallmanzig, J. (1991) Diagnosis study of the St. Marienkirche, Lippstadt, a mediaeval greensandstone monument, for conservatory planning. In: *"Science, Technology and European Cultural Heritage"*, Butterworth Heinemann, Oxford, UK, pp. 939-942.

[86] Wüst, R. A. J., Schlüchter, C. (2000) The Origin of Soluble Salts in Rocks of the Thebes Mountains, Egypt: The Damage Potential to Ancient Egyptian Wall Art. *Journal of Archaeological Science* 27:1161-1172. Doi:10.1006/jasc.1999.0550

[87] Cooper, P., Dowding, P., Lewis, J. O., Mulvin, L., O'Brien, P., Olley, J., O'Daly, G. (1991) Contribution of calcium from limestone and mortar to the decay of granite walling. In: *"Science, Technology and European Cultural Heritage"*, Butterworth Heinemann, Oxford, UK, pp. 456-461.

[88] Smith, B. J., Turkington, A. V., Warke, P. A., Basheer, P. A. M., McAlister, J. J., Meneely, J., Curran, J. M. (2002) Modelling the rapid retreat of building sandstones: a case study from a polluted maritime environment. In: *"Natural Stone, Weathering Phenomena, Conservation*

[109] Thomachot, C., Jeannette, D. (2004) Effects of iron black varnish on petrophysical properties of building sandstone. *Environmental Geology* 47:119-131. Doi:10.1007/s00254-004-1139-4

[110] Rescic, S., Fratini, F., Tiano, P. (2007) On site evaluation of "mechanical-physical" properties of the Maastricht limestone. STONE - Newsletter on stone decay 2:38-39. Freely available (May 2012) at http://www.qub.ac.uk/geomaterials/weathering/newsletter/issue_2.pdf

[111] Sabbioni, C. (2003) Mechanisms of air pollution damage to stone. In: *"The Effects of air pollution on the built environment".* Imperial College Press, London, pp. 63–106.

[112] Brimblecombe, P., Grossi, C. M. (2010) Potential damage to modern building materials from 21st century air pollution. *Scientific World Journal* 10:116-25.

[113] Cardell, C., Benavente, D., Rodríguez-Gordillo, J. (2008). Weathering of limestone building material by mixed sulfate solutions. Characterization of stone microstructure, reaction products and decay forms. *Materials Characterization* 59:1371-1385. Doi:10.1016/j.matchar.2007.12.003

[114] Yu, S., Oguchi, C. T. (2009), Complex relationships between salt type and rock properties in a durability experiment of multiple salt–rock treatments. *Earth Surfaces Processes Landforms* 34:2096–2110. Doi:10.1002/esp.1904

[115] Nord, A. G., Svärdh, A., Tronner, K. (1994) Air pollution levels reflected in deposits on building stone, *Atmospheric Environment* 28:2615-2622. Doi:10.1016/1352-2310(94)90435-9

[116] Mottershead, D., Gorbushina, A., Lucas, G., Wright, J. (2003) The influence of marine salts, aspect and microbes in the weathering of sandstone in two historic structures. *Building and Environment* 38:1193-1204. Doi:10.1016/S0360-1323(03)00071-4

[117] Stefanis, N-A., Theoulakis, P., Pilinis, C. (2009) Dry deposition effect of marine aerosol to the building stone of the medieval city of Rhodes, Greece. *Building and Environment* 44:260-270. Doi:10.1016/j.buildenv.2008.03.001

[118] Costa, E. A. L., Campos, V. P., Silva Filho, L. C. P., Greven, H. A. (2009) Evaluation of the aggressive potential of marine chloride and sulfate salts on mortars applied as renders in the Metropolitan Region of Salvador – Bahia, Brazil. *J. Environ. Management*. 90:1060-1068. Doi:10.1016/j.jenvman.20 08.0 4.0 06

[119] Pilkey, O. H., Pilkey-Jarvis, L. (2007) *Useless Arithmetic: Why Environmental Scientists Can't Predict the Future*. Columbia University Press, New York.

In: Advances in Geochemistry Research
Editor: Jorge Sanjurjo-Sánchez ISBN: 978-1-62618-245-5

Chapter 4

GENERATION OF HIGH-K GRANITES AT THE ARCHAEAN – PROTEROZOIC BOUNDARY

Shukrani Manya*
Department of Geology, University of Dar es Salaam,
Dar es Salaam, Tanzania

ABSTRACT

The Archaean crustal provinces of the world are dominated by the rocks of the Tonalite-Trondhjemite-Granodiorite (TTG) series and account for 50% of the exposed Archaean crust. Towards the end of Archaean, however, TTG magmatism was replaced by the generation of high-K granites which continued into the Proterozoic and Phanerozoic. The late Archaean (~2.65 Ga) high-K granites of the Tanzania Craton some of which intrude the greenstone sequence of northern Tanzania are compared with the early Proterozoic (~ 1.92 – 1.96 Ga) high-K granites from the Ubendian belt of southwestern Tanzania. Granites from both late Archaean and early Proterozoic crustal terranes are pink in colour, and are composed of K-feldspar, plagioclase, quartz, biotite, less hornblende with minor phases including sphene, zircon and chlorite. It is in the Proterozoic granites that muscovite and tourmaline become important phases.

* Email: shukrani73@udsm.ac.tz or shukrani73@yahoo.com.

The compositional similarities of the high-K granites from both late Archaean and early Proterozoic terranes include high SiO_2 (> 70 wt %), high K_2O (≥ 4.0 wt %) which leads to K_2O/Na_2O ratio of > 1, Al_2O_3 contents lower than those of TTG (< 15 wt %) and low MgO < 1.2 wt % and Ni (< 20 ppm). Trace element compositions reveal that, the high-K granites from both crustal terranes show comparable concentrations of Y, MREE and HREE, and display the negative Eu anomaly (Eu/Eu* = 0.5 – 0.6 on average).

The major compositional differences between them are discerned in the most incompatible elements. High-K granites from the Proterozoic terrane consistently show higher concentrations of Zr, Hf, Th, Ba, Sr, K and ∑REE than their late Archaean counterparts. On the basis of the observed compositional similarities and differences, the key trace element ratios employed to distinguish the two magmas involve an incompatible element and middle to heavy REE (i.e. Zr/Sm, Zr/Y, Zr/Nb, Ba/Nb, Th/Ta) as well as the K_2O/Na_2O ratio. These ratios are a reflection of the primary petrogenetic feature inherent from the source rocks. The field relationships and geochemical features of the high-K granites from both crustal terranes are consistent with their derivation from partial melting at pressures and depths where plagioclase was a residual phase. Whereas the generation of the high-K granites in the Archaean would require partial melting of the juvenile igneous rocks including the TTG and greenstones that were emplaced in previous 10 to 20 Ma, those in the Proterozoic would require partial melting of the more evolved crustal rocks including the Archaean rocks. The generation of high-K granitic magmas at the Archaean-Proterozoic boundary marked a transition from growth of the Neoarchaean continental crust through the addition of juvenile mantle-derived material to intra-crustal recycling of pre-existing material. It also marked the final stages of the stabilization of the Archaean Cratons worldwide.

Keywords: High-K granites, late Archaean, early Proterozoic, Tanzania Craton

INTRODUCTION

The Archaean crustal provinces of the world are dominated by the rocks of the Tonalite-Trondhjemite-Granodiorite (TTG) series and account for more than 50 % of the exposed Archaean crust [1]. In many cratons, the TTG constitute the gneissic basement to the greenstones [2], although they may also occur as individual syn- to post-tectonic plutons [3]. Towards the end of Archaean, the TTG magmatism and greenstone belt volcanism was replaced by high-K granites (and Mg-rich granodiorites) and Condie [1] estimated that they constitute 20% of the exposed crust in most Archaean cratons. Exceptions are noted in some few cratons and granite-greenstone terranes (e.g. the Yilgarn block of Western Australia [4]) where high-K granites are much more voluminous than the TTG gneisses and plutons. Typology of late Archaean granites based on Dharwar craton showed that the high-K granites (also known as biotite-bearing granites) are accompanied by occasional leucogranites and peraluminous granites as well as peralkaline granites in few cratons [5].

The high-K granitic magmatism appear to have started in late Archaean and continued into the early Proterozoic through to the Phanerozoic [6, 7]. The generation of these potassic magmas at the Archaean – Proterozoic (AP) boundary has been interpreted to indicate that mid- to shallow level intracrustal melting became significantly more important after the Archaean. This is best reflected in the geochemistry of the high-K granites in which the near ubiquity of negative europium anomalies and flat HREE patterns indicate that magma generation occurred at relatively shallow, mid-crustal depths with plagioclase, rather than garnet or amphibole, was a residual or fractionating phase [6]. In this chapter, the geochemical characteristics of the high-K granites from the late Archaean Musoma-Mara greenstone belt of Tanzania Craton and the early Proterozoic Ubendian belt are examined in order to draw their compositional similarities and differences. Distinguishing geochemical features are proposed and constraints on their genesis are made.

Geological Setting of the Samples

The Archaean high-K granites presented in this chapter are from the Musoma-Mara greenstone belt (MMGB) of northern Tanzania (Figure 1). The MMGB is a Neoarchaean granite-greenstone terrane forming part of the Tanzania Craton. The greenstone sequence in the MMGB is made of 2676 – 2669 Ma high-Mg andesites that are overlain by ~2668 Ma felsic volcanic rocks of adakitic composition [9]. These greenstones are intruded by syn-orogenic granitoids of TTG composition, which are considered to represent additions of juvenile mantle-derived material into the continental crust at ~2668 Ma [9]. The emplacement of the high-K granites at ~2649 Ma [9] concluded the magmatic evolution of the MMGB. Unlike in other greenstone belts of the world where TTG are more abundant than the high-K granites [4]; Manya et al. [10] and Mshiu and Maboko [11] observed that the high-K granites are the most abundant rocks in the MMGB and occupy 60% of all rocks in the area. These granites are composed of plagioclase + K-feldspar + quartz + biotite ± sphene ± zircon ± hornblende ± chlorite. On the other hand, the Proterozoic high-K granites presented here were sampled on an almost linear traverse from central Tanzania within the Tanzania Craton, south-westward into the early Proterozoic Ubendian belt (Figure 1). The 2.0 Ga Ubendian belt [12, 13] is a NW-SW trending belt that boarders the Tanzania Craton to the west and southwest. The Ubendian belt is subdivided into eight distinct structural blocks or terranes persistently trending NW- SE each separated by ductile shear zone up to 600 km in length; with most shear zones showing a NW- dextral strike-slip displacement [14, 15]. These terranes include Ubende, Wakole, Katuma, Ufipa, Upangwa, Nyika, Mbozi and Lupa.

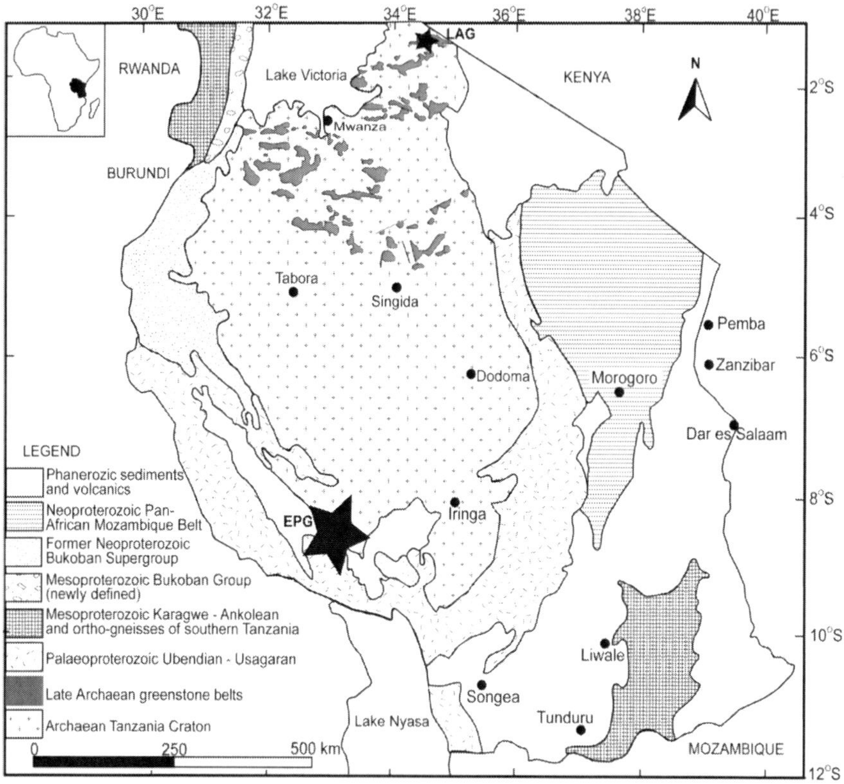

Figure 1. Geological sketch map of Tanzania (modified after Pinna et al. [8]) showing the location areas for the studied granites indicated by a black star. LAG = Late Archaean high-K Granites from the Musoma-Mara greenstone Belt reported in Manya et al. [10] and EPG = Early Proterozoic high-K Granites reported in Manya [20].

The detailed descriptions of the Ubendian belt are reported by Smirnov et al. [16] and Nanyaro [17] and their structural analyses of the gneissic fabric and mineral lineations are given by Theunissen et al. [18] and Boven et al. [19]. The Ubendian Belt consists of high-grade metamorphic rocks of sedimentary and igneous origin, including granulites, migmatite, amphibolites, biotite gneisses, schists, quartzites, eclogites and marble. Granites are the main type of intrusive rocks, but there are also granodiorites, diorites, gabbro and carbonatite rocks [14]. The granites of the area from which the samples of this study were taken are divided into three types: the porhyritic biotite granites, the coarse grained Saza and Chunya granites, and the Ilunga granites which are medium to coarse grained aplogranites. The last two are named according to their local places of occurrences. These rocks are composed of alkali feldspar (microcline) with a characteristic pinkish colour, quartz, biotite, with few specimens composed of muscovite and tourmaline. Minor phases include hornblende, zircon, sphene, chlorite. These granites yielded Nd-model ages of 2.4 – 2.0 Ga [20] and Mnali [21] reported zircon U-Pb age of 1920 – 1960 Ma and interpreted them as being emplacement ages for the Saza granodiorites and Ilunga granites. Granitic rocks in other parts of the Ubendian belt for which U-Pb zircon ages are available cluster around ages of 2.0 – 1.8 Ga (i.e. the Mbarali granite (2026 ± 8 Ma), the Ufipa granite-gneiss (1864 ± 32 Ma)) with a few reported to have formed at 0.7 Ga (e.g. the Kalunduru granite (724 ± 6 Ma)) [12, 19]. Previous workers in the Ubendian [12, 19] regarded the 2.0 – 1.8 Ga high-grade metamorphism as the only tectonothermal event to have affected the Ubendian – Usagaran orogen. However, more recently, Nelson et al. [22] reported three tectonothermal events that have affected the Ubendian orogen: These include Paleoproterozoic (Eburnean) ocean floor subduction at 1890 – 1860 Ma, which was followed by Mesoproterozoic (Kibaran) 1170 – 1010 Ma crustal thickening and granulite facies mylonitic reworking of the Paleoproterozoic rocks and lastly Neoproterozoic convergent margin at 590 – 520 Ma and subsequent eclogite-facies metamorphism.

GEOCHEMISTRY

The chemical composition of the high-K granites from both the late Archean and early Proterozoic crustal terranes are presented in Table 1 and in Figures 2 – 7. These rocks are characterized by high contents of SiO_2 (> 70 wt % on average), K_2O (> 4 wt %) and have K_2O/Na_2O > 1.0 (Table 1). On Total Alkali Silica diagram of Cox et al. [23] (Figure 2), they both plot straddling the boundary between sub-alkali granite to alkali granite with the Proterozoic granites tending to higher total alkalis ($Na_2O + K_2O \geq 10$ wt %). The tendency to higher alkalis owing to the higher contents of K_2O is further shown in the triangular Na_2O-K_2O-CaO diagram (Figure 3) in which both the late Archaean and Proterozoic granites plot in the field for biotite-granites of Moyen et al. [5]. The diagram by Moyen et al. [5] is used to differentiate various types of the Late Archean granitoids and the plotting of both the late Archaean and early Proterozoic high-K granites in the field of biotite granites corroborate their similar geochemical features at least in terms of their alkali contents.

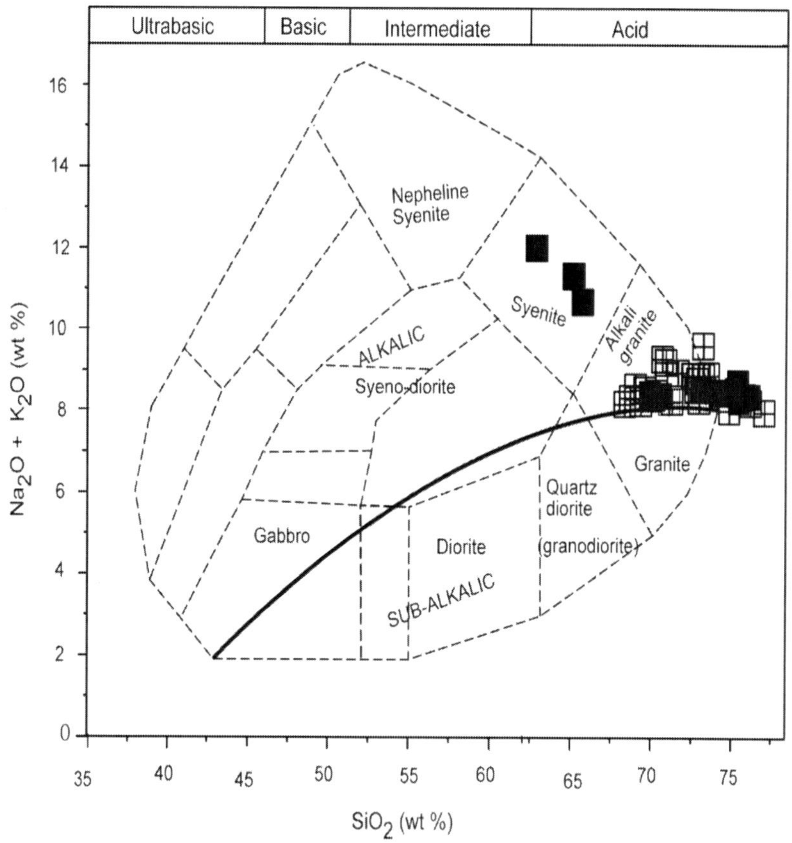

Figure 2. Classification of the studied high-K granites using the Total Alkali Silica (TAS) diagram of Cox et al. [23].

Table 1. Major (wt %) and trace (ppm) element data for the high-K granites from the late Archaean Tanzania craton and early Proterozoic Ubendian belt of southwest Tanzania

Late Archaean high-K granites				Early Proterozoic high-K granites		
	Average	Standard	Range	Average	Standard	Range
	(n = 29)	deviation		(n = 12)	deviation	
SiO_2	72.9	2.34	68.9 - 77.8	71.4	4.7	63.3 - 76.8
TiO_2	0.2	0.12	0.01 - 0.54	0.27	0.16	0.07 - 0.54
Al_2O_3	14.0	0.82	11.8 - 15.1	14.1	1.6	11.8 - 17.7
Fe_2O_3	1.81	0.82	0.29 - 3.56	2.21	1.02	0.76 - 3.98
MnO	0.05	0.02	0.01 - 0.11	0.05	0.01	0.02 - 0.06
MgO	0.33	0.25	0.03 - 0.99	0.47	0.36	0.08 - 1.18
CaO	1.21	0.55	0.22 - 2.41	1.12	0.57	0.31 - 2.05
Na_2O	3.96	0.47	3.27 - 5.70	3.89	0.80	2.78 - 5.85
K_2O	4.59	0.42	3.71 - 5.44	5.31	0.81	4.18 - 6.82
P_2O_5	0.06	0.05	0.004 - 0.19	0.10	0.10	0.02 - 0.34
LOI	0.68	0.18	0.35 - 0.99	0.57	0.20	0.32 - 0.98
Total	99.8	0.24		99.5	0.55	
Ni	< 20		< 20	< 20		
Zr	161	81.2	20 - 401	302	259	57 - 936
Hf	4.88	1.72	0.69 - 9.32	8.95	6.02	3.50 - 23.6
Nb	13.0	9.95	1.35 - 57.4	12.6	3.75	8.0 - 21
Ta	3.89	11.45	0.15 - 60.1	0.96	0.36	0.50 - 1.70
Rb	312	139	188 - 880	200	45	128 - 268
Sr	224	115	15 - 412	276	322	40 - 1208
Y	22	12	2.31 - 43.1	18.3	12.8	7.0 - 53
Ba	569	328	15 - 1034	1128	1256	184 - 4758

Table 1. (Continued)

Late Archaean high-K granites				Early Proterozoic high-K granites		
	Average	Standard	Range	Average	Standard	Range
	(n = 29)	deviation		(n = 12)	deviation	
La	45.9	25.6	4.71 - 96.7	57.5	32.0	8.9 - 127
Ce	88	49	12 - 189	111	56	19 - 223
Pr	9.60	5.27	1.30 - 20.8	12.4	6.1	1.92 - 22.3
Nd	33.3	17.9	4.70 - 73.6	38.2	16.8	5.51 - 58.3
Sm	5.34	2.62	0.72 - 11.4	6.38	2.9	0.90 - 9.70
Eu	0.77	0.42	0.05 - 1.80	1.09	0.7	0.15 - 2.65
Gd	4.32	2.15	0.66 - 9.59	4.46	2.0	0.60 - 6.90
Tb	0.60	0.30	0.07 - 1.26	0.71	0.3	0.31 - 1.31
Dy	3.13	1.56	0.34 - 6.20	3.53	2.2	0.62 - 8.92
Ho	0.62	0.31	0.07 - 1.22	0.64	0.5	0.11 - 1.90
Er	1.64	0.80	0.18 - 3.09	1.91	1.5	0.41 - 6.14
Tm	0.27	0.12	0.03 - 0.49	0.29	0.2	0.07 - 0.99
Yb	1.88	0.83	0.21 - 3.52	1.94	1.6	0.60 - 6.51
Lu	0.28	0.12	0.04 - 0.51	0.29	0.2	0.10 - 0.91
Th	25.3	14.87	5.70 - 68.5	30.8	21.0	11.3 - 87.1
Pb	31.7	5.77	23.9 - 47.0	31.3	8.6	19 - 43
U	6.75	3.07	2.60 - 14.8	4.03	1.6	2.30 - 7.40
La/Yb	26.5	12.5	1.80 - 61.3	37.3	23.6	4.90 - 97.7
La/Yb_{CN}	17.9	8.42	1.20 - 41.3	26.7	17.0	3.50 - 70.1
Eu/Eu*	0.54	0.25	0.04 - 1.27	0.63	0.3	0.08 - 1.05
K_2O/Na_2O	1.18	0.20	0.68 - 1.60	1.40	0.3	1.03 - 2.03
Zr/Y	8.74	4.22	0.62 - 20.1	21.3	23.2	2.75 - 85.1

Late Archaean high-K granites				Early Proterozoic high-K granites		
	Average (n = 29)	Standard deviation	Range	Average (n = 12)	Standard deviation	Range
Zr/Sm	32.0	11.1	5.86 - 54.7	45.9	24.9	22.8 - 110
Th/Ta	24.8	19.3	0.09 - 108	41.2	47.1	14.1 - 174
Zr/Nb	15.4	7.29	0.43 - 35.9	24.3	20.9	6.33 - 72.1
Ba/Nb	59.0	39.3	1.01 - 134	98.3	128	13.1 - 458
REE total	196	104	29.7 - 417	240	115	39.6 - 451

Late Archaean high-K granites are from Manya et al. [10] whereas early Proterozoic high-Kgranites are from Manya [20].

The high-K granites from both crustal terranes show Al_2O_3 contents of < 15 wt %, MgO < 1.2 wt %, Ni < 20 ppm, Sr < 500 ppm (with an outlier) and these compositions are in stark contrast to those of the Archaean TTG which according to Martin [25], exhibit higher contents of Al_2O_3 (> 15 wt %), Ni ≥ 24 ppm and Sr (> 500 ppm) (Table 1, Figure 4). They are also dissimilar to the A-type granites as well as the two mica granites of Moyen et al. [5].

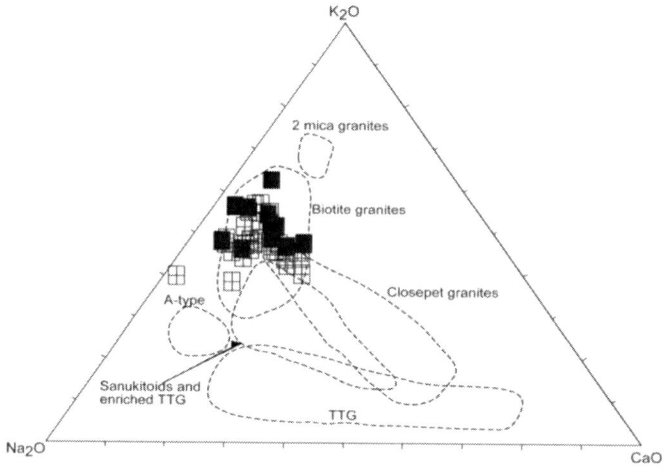

Figure 3. Triangular Na2O-K2O-CaO diagram for the studied high-K granites. Classification Fields adopted from the worldwide Late Archaean granitoids reported by Moyen et al. [5].

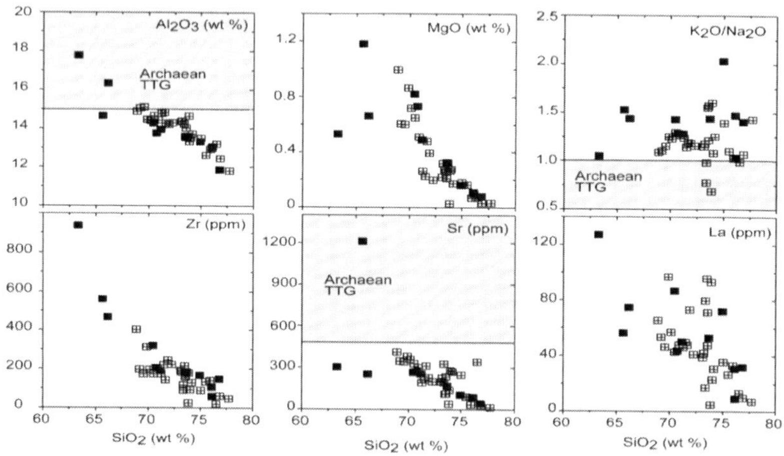

Figure 4. Major and trace element variation diagrams for the studied high-K granites.

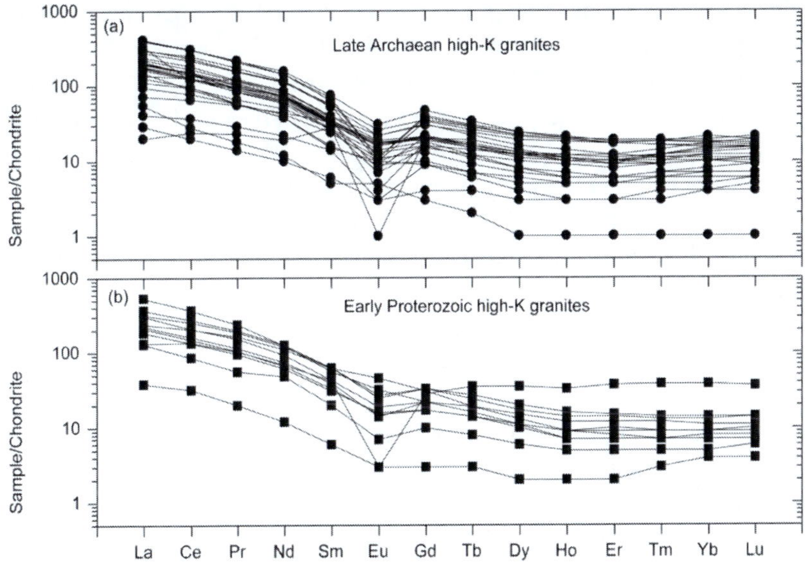

Figure 5. Chondrite normalized REE patterns for (a) Late Archaean high-K granites (b) Early Proterozoic high-K granites. Normalizing values after Sun and McDonough [24].

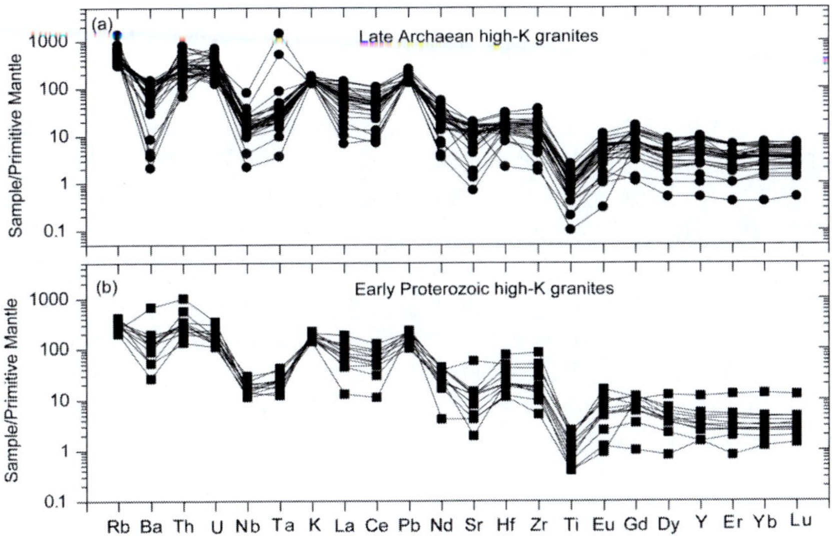

Figure 6. Primitive mantle normalized multi-element diagram for (a) Late Archaean high-K granites (b) Early Proterozoic high-K granites. Normalizing values after Sun and McDonough [24].

On chondrite normalized rare earth elements (REE) diagrams (Figure 5), high-K granites from both crustal terranes show fractionated patterns with enrichment of the light REE (LREE) over the heavy REE (HREE) (La/Yb_{CN} = 17.9 and 26.7 on average for late Archaean and early Proterozoic rocks, respectively).

The rocks exhibit characteristic negative Eu anomalies (Eu/Eu^* = 0.5 and 0.6, on average, for late Archean and early Proterozoic granites, respectively). The total REE contents are, however, relatively higher for early Proterozoic (ΣREE = 240 ppm on average) than their late Archaean counterparts (ΣREE = 196 ppm on average) although the two rock suites have comparable MREE and HREE.

On primitive mantle normalized multi-element diagrams (Figure 6), the high-K granites from both crustal terranes show enrichment of the incompatible elements like Rb, K, Th, and Pb relative to adjacent elements and negative anomalies of Nb, Ta, Ti and Sr. The notable feature is the relative enrichment of Ba, Th, K, Zr and Hf for the high-K granites of the early Proterozoic compared to their late Archaean counterparts (Table 1, Figures 4 and 6).

DISCUSSION

The geochemical characteristics of the high-K granites from the late Archaean and early Proterozoic crustal terranes of Tanzania provide an opportunity for their comparison and contrast as well as constraining their petrogenesis. They both exhibit high contents of SiO_2 (> 70 wt %), total alkalis (8.6 and 9.19 on average for Archaean and Proterozoic rocks, tespectively), low Al_2O_3 (generally < 15 wt %), MgO (< 1.2 wt %), Ni (< 20 ppm), negative anomalies of Nb, Ta, Ti, and Sr, as well as an overall enrichment of the incompatible elements. These features are comparable to those of biotite-granites in the classification of late Archaean granitoids of Moyen et al. [5] which are in stark contrast to the late Archaean TTG, two mica granites and A-type granites.

The high-K granites from the early Proterozoic are characterized by higher contents of incompatible elements like Ba (up to 4758 ppm), Sr, Th, K, Zr and Hf as well as higher total REE compared to their late Archaean counterparts. This provides a point of departure for which the early Proterozoic high-K granites can be distinguished from those of the late Archaean. As shown in Figure 7, the key trace elements that can distinguish the two closely similar magmas involve a ratio between an incompatible element and a middle or heavy REE.

Thus, ratios of K_2O/Na_2O, Zr/Sm, Zr/Y, Zr/Nb, Ba/Nb, Th/Ta are proposed as key distinguishing geochemical features of the two high-K magmas generated at the A-P boundary. The early Proterozoic magmas are thus exhibiting higher K_2O/Na_2O, Zr/Sm, Zr/Y, Zr/Nb, Ba/Nb and Th/Ta ratios than their late Archaean equivalents (Figure 7) and is a reflection of the primary petrogenetic feature inherent from the source rocks.

The geochemical features of both magmas at the A-P boundary provide important clues to their petrogenesis. The low contents of MgO, Ni and Fe_2O_3 argue against the derivation of these magmas directly from the mantle. The mantle derived rocks do equilibrate with peridotite resulting in higher MgO and Ni contents [25] and as such, the high-K granites from both the Archaean and Proterozoic crustal terranes are crustal derivatives.

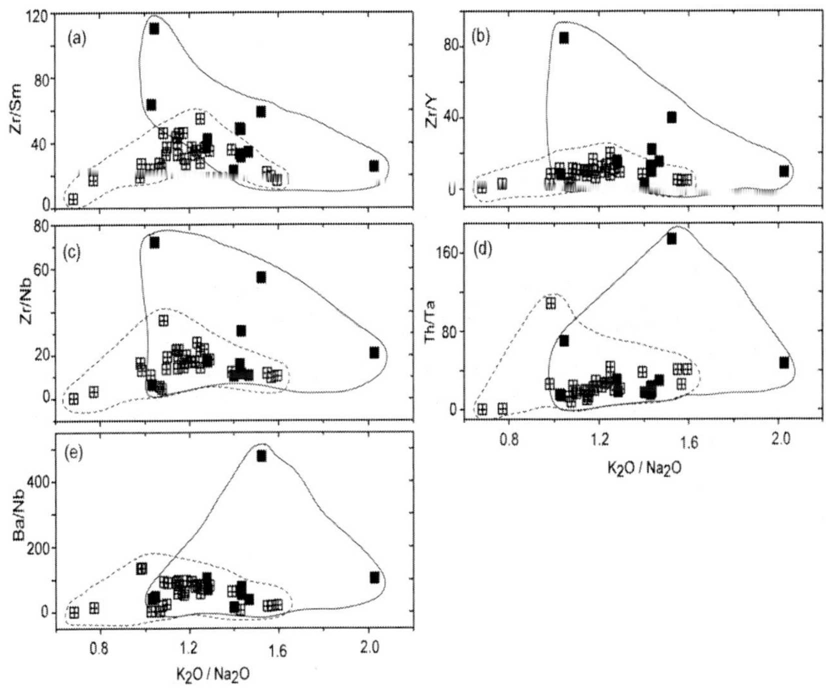

Figure 7. K_2O/Na_2O versus (a) Zr/Sm (b) Zr/Y (c) Zr/Nb (d) Th/Ta (e) Ba/Nb diagrams used to distinguish high-K granites of the Late Archaean from those of the Early Proterozoic.

The negative Eu anomalies of the rocks (Eu/Eu* = 0.5 – 0.6 on average) coupled by their low contents of Sr (< 500 ppm, albeit with one outlier), Al_2O_3 and CaO, together suggest the presence of plagioclase as a residual phase during partial melting of their source (although this can also be caused by fractional crystallization of plagioclase during magmatic differentiation).

The generation of such large terranes of granites would require partial melting over large crustal areas (rather than fractional crystallization) and partial melting occurred at low pressure conditions and within the crust where plagioclase was a stable phase. For late Archaean crustal terranes where Sm-Nd isotopic data is available, it is shown that εNd values for the high-K granites are similar to those of the preceding TTG and greenstones [10, 26] indicating that their source did not have long crustal residence time. This has been interpreted as reflecting the derivation of the late Archaean high-K granites through partial melting of the pre-existing TTG and greenstones which were emplaced in the previous 10 to 20 Ma. This being the case, the negative Nb, Ta and Ti anomalies observed in the high-K granites mirrors their source rocks and should not have any reflection on the subduction-related affinity which is mainly attributed to rocks with such geochemical features [27].

The higher contents of incompatible elements in the early Proterozoic high-K granites than their late Archaen equivalents is interpreted to reflect crustal anatexis of the more evolved rocks including the late Archaean rocks. Most Proterozoic granites show Nd model ages of up to 500 Ma older than their emplacement ages as well as slightly negative εNd values suggesting longer crustal residence time [28]. In particular, Maboko and Nakamura [28] demonstrated that the early Proterozoic granites of southeastern Tanzania represented juvenile mantle material that was extracted from the mantle at about 2000 Ma and subsequently assimilated with Archaean crustal component from the adjacent Tanzania Craton. Crustal anatexis involving late Archaean rocks is consistent with the high-K granites of the early Proterozoic exhibiting enrichment of incompatible elements like Ba, Rb, Th, Zr, Hf and K.

The A-P boundary is of uttermost importance in the geological history of our planet. The majority of the rocks generated prior to the A-P boundary include TTG, greenstones and sanukitoids [5]. These rocks, despite their ever ongoing debate on their geodynamic setting, are considered by many researchers to have been derived from partial melting of the hydrated basalt, most likely in subduction zones [27, 29]. This encompasses the addition of juvenile mantle derived material to the crust. On the contrary, the generation of high-K granitic magmas at the Archaean-Proterozoic boundary marked a transition from growth of the Neoarchaean continental crust through the addition of juvenile mantle-derived material to intra-crustal recycling of pre-existing material [7]. It also marked the final stages of the stabilization of the Archaean Cratons worldwide.

ACKNOWLEDGMENTS

I would like to acknowledge the financial support I have received from Sida/SAREC over the years which has facilitated fieldwork in the Archaean greenstone belts and Proterozic crustal terranes of Tanzania, as well as the analyses of the geochemical samples. The editorial handling of the manuscript by Jorge Sanjurjo Sánchez is highly acknowledged.

REFERENCES

[1] K. C. Condie, *Chem. Geol.* 104, 1-37 (1993).

[2] B. F. Windley, *The Evolving Continets*, John Wiley and Sons, Chichesster (1995).

[3] D. C. Champion and R.H. Smithies, *Archaean granites of the Yilgarn and Pilbara cratons, Western Australia: secular changes* (1999). B. Barbarin, –The origin of granites and related rocks", 4th Hutton symposium Abstracts Doc. BRGM 290, pp. 137.

[4] P. J. Sylvester, *Archaean granite plutons* (1994). K. C. Condie, –Archaean Crustal Evolution," Elsevier, Amsterdam, pp. 261-314.

[5] J.-F. Moyen, H. Martin, M. Jayananda and B. Auvray, *Precam. Res.* 127, 103-123 (2003).

[6] K. C. Condie, *Plate Tectonics and crustal evolution*, Butterworth Heinemann, 282, (1997).

[7] I. S. Kemp and C. J. Hawkesworth, *Granitic perspectives on the generation and secular evolution of the continental crust* (2003). R.L. Rudnick, –Treatise on Geochemistry", Elsevier – Pergamon, Oxford, pp. 349 – 410.

[8] P. Pinna, S. Muhongo, A. Mcharo, E. Le Goff, Y. Deschamps, F. Ralay and J. P. Milesi, *Geology and mineral map of Tanzania*, Geological Survey of Tanzania, Dodoma (2008).

[9] S. Manya, K. Kobayashi, M. A. H. Maboko and E. Nakamura, *J. Afr. Earth Sci.* 45, 355 - 366 (2006).

[10] S. Manya, M. A. H. Maboko and E. Nakamura, *Precam. Res.* 159, 231 - 240, (2007).

[11] E. E. Mshiu and M. A. H. Maboko, *J. Afr. Earth Sci.* 66-67, 1 - 12 (2012).

[12] J. L. Lennoir, J. P. Liegeois and J. Klerkx, *J. Afr. Earth Sci.* 19, 169 - 184 (1994).

[13] Möller, P. Appel, K. Mezger and V. Schenk, Geology, 23, 1067 - 1070 (1995).

[14] R. McConnell, *Geological Survey of Tanganyika Bulletin* 19, 62 (1950).

[15] J. Sutton, J. Watson, and T. James, *Geological Survey of Tanganyika Bulletin* 22, (1954).

[16] V. Smirnov, V. Pentelkov, V. Tolochko, M. Trifan and S. Zhukov, *Geology and Minerals of the Central Part of the Western Rift, Tech. Rep. Mineral and Resource Division, Dodoma, Tanzania*, (1973).

[17] J. T. Nanyaro, N. K. Basu, S. M. Muhongo, A. H. Mruma, S. A. Djare, I. Mduma, K. M. Mudiguza, S. P. Van and J. Klerkx, *Rapport Annuel – Musee Royal de l'Afrique Centrale*, 147–152 (1982).

[18] K. Theunissen, J. Klerkx, A. Melnikov and A. H. Mruma, *Tectonics* 15, 776 - 790 (1996).

[19] Boven, K. Theunissen, E. Sklyarov, J. Klerkx, A. Melnikov, A. Mruma and L. Punzalan, *Precam Res.* 93, 119 – 137 (1999).

[20] S. Manya, Gondwana Res. 20, 325 – 334 (2011).

[21] S.R. Mnali, ―Palaeoproterozoic felsic magmatism and associated gold-quartz vein mineralization in western part of the Lupa Gold Field, south-western Tanzania" *PhD Thesis, University of Dar es Salaam,* 198 (1999).

[22] N. Boniface, V. Schenk and P. Appel, *Precam. Res.* 192 – 195, 16 – 33 (2012).

[23] K. G. Cox, J. D. Bell and R.J. Pankhurst, The interpretation of igneous rocks, George, Allen and Unwin, London (1979).

[24] S. S. Sun and W. F. McDonough, *Chemical and isotopic systematic of oceanic basalts: implication for mantle composition and processes.* A. D. Saunders, and M. J. Norry, ―Magmatic in Ocean Basins". *Geological Society of London Special Publication* 42, 313-345 (1989).

[25] H. Martin, *Lithos* 46, 411 – 429 (1999).

[26] M. Jayananda, J.-F. Moyen, H. Martin, J.-J. Peucat, B. Auvray and B. Mahabaleswar, B., *Precam. Res.* 99, 225 – 254 (2000).

[27] H. Martin, R. H. Smithies, R. Rapp, J.-F. Moyen and D. Champion, *Lithos* 79, 1-24 (2005).
[28] M. A. H. Maboko and E. Nakamura, *Precam. Res.* 77, 105 – 115 (1996).
[29] M. S. Drummond and M.J. Defant, *J. Geophy. Res.* 95B, 21503 – 21521 (1990).

In: Advances in Geochemistry Research
Editor: Jorge Sanjurjo-Sánchez
ISBN: 978-1-62618-245-5

Chapter 5

BIOMINERALIZATION BY CAVE DWELLING MICROORGANISMS

A. Z. Miller[1*], A. Dionísio[1], V. Jurado[2], S. Cuezva[3], S. Sanchez-Moral[4], J. C. Cañaveras[3] and C. Saiz-Jimenez[2]

[1] Centro de Petrologia e Geoquímica, Instituto Superior Técnico, Lisboa, Portugal

[2] Instituto de Recursos Naturales y Agrobiología de Sevilla, IRNAS-CSIC, Sevilla, Spain

[3] Laboratorio de Petrología Aplicada, Departamento de Ciencias de la Tierra y del Medio Ambiente, Universidad de Alicante, Alicante, Spain

[4] Museo Nacional de Ciencias Naturales, MNCN-CSIC, José Gutiérrez Madrid, Spain

ABSTRACT

Microorganisms inhabit all possible biosphere ecosystems including subsurface rock environments, such as caves, mines, tunnels, etc. These environments are nutrient-limited and contain a wide variety of redox interfaces and microorganisms able to promote biomineralization processes. By interacting with minerals, microorganisms play an important role in reshaping the mineral environment, and may help to

* E-mail address: ana.miller@ist.utl.pt.

form features such as mineral deposits and some types of speleothems in cave environments.

Biomineralization, the processes by which organisms form minerals, is a fascinating topic that is receiving renewed attention by various fields of science. This chapter presents an overview of the microbial mineralization in caves and briefly describes which minerals are most commonly precipitated in the subsurface environments.

INTRODUCTION

Caves are windows into the subsurface which allow scientists to explore the space below the Earth's surface. Most common types of caves are those formed in limestone and other calcareous rocks, as well as lava tubes in basaltic rocks. Remaining types include those formed in gypsum, quartzite, sandstone and granite, which are usually limited in extent and, in some cases, rarely found and investigated. Nevertheless, most types of rocks are penetrated by humans for a variety of purposes such as mining for metals, extraction of groundwater and waste hard rock repositories, providing underground research facility (Pedersen et al., 1996; Foster and Ashley, 2002; Anderson et al., 2006).

The underground environment and its interior atmosphere are a highly complex environment separated from the exterior atmosphere by two membranes or isolating layers with very different geochemical and geophysical characteristics: the soil in direct connection with the external atmosphere, and the rock (with its porous network and discontinuities) in direct contact with the interior atmosphere. These characteristics together with the infiltration of waters through the rocky massif discontinuities provide specific environmental conditions which tend to be stable the year round. Despite the moderate temperature and high humidity, caves are considered extreme environments due to their limited amounts of surface-energy inputs, mineral concentrations and absence of sunlight, confining life for highly specialized microorganisms and adapted to starvation conditions (Schabereiter-Gurtner et al., 2004; Barton, 2006).

The evidence for extant microbial activity in caves became more compelling with the discovery of mineral deposits difficult to explain through purely geologic or inorganic processes. Several studies suggested that calcium carbonate deposits known as moonmilk, pool finger precipitation and ferromanganese deposits, usually considered to be abiotic in composition and genesis, are indeed biologic in origin arising from important geomicrobiological interactions that underpin energy conservation and life in subsurface environments (Barton et al., 2001; Melim et al., 2001; Northup et al., 2003; Spilde et al., 2005; Cañaveras et al., 2006). By interacting with minerals, microorganisms play an important role in reshaping the mineral environment, and may help to form features such as mineral deposits and some types of speleothems.

Microorganisms, inhabiting all possible biosphere ecosystems including caves, are able to promote mineral biotransformation and biomineralization processes, which have been carried out by organisms over Earth history (Hofmann et al., 2008). In addition, the interwined relations of biogenic minerals, or biominerals, to earth environments are fundamental components of natural biogeochemical cycles (Nealson and Stahl, 1997; Newman and Banfield, 2002; Van Cappellen, 2003). Despite numerous studies performed to determine the mechanisms of biomineralization, the precise biomineral formation mechanisms remain uncertain.

This chapter presents an overview of the microbial mineralization in caves and briefly describes which minerals are most commonly precipitated in the subsurface environments.

BIOMINERALIZATION

Biomineralization is the natural processes by which living organisms form minerals from bioorganic molecules and inorganic solids (Bäuerlein, 2004). Most biominerals, formed under controlled conditions, are composite products since they comprise both an organic matrix and nano- or micro-scale amorphous or crystalline minerals. They are the products of the selective uptake of elements from the surrounding environment and their incorporation into functional structures under strict biological control (Mann, 2001). The living organism provides a chemical environment that controls the nucleation and growth of the mineral phases. Biominerals meet the criteria for being true minerals however they are often distinguishable from their inorganically produce counterparts by their peculiar shape, size, crystallinity, isotopic and trace element compositions (Weiner and Dove, 2003).

Until the early 1980s the field was known as ―calcification," reflecting the predominance of biologically formed calcium-containing minerals. An expected result since calcium fulfills many fundamental functions in cellular metabolism. As more and more biogenic minerals were discovered containing other cations, the field became known as ―biomineralization." (Weiner and Dove, 2003).

Organisms which have the capability of precipitating minerals are present in all taxonomic groups, from Bacteria to Chordata, being animals the foremost group able to induce the precipitation of minerals (Lowenstam, 1981; Jimenez-Lopez et al., 2007).

According to Benzerara et al. (2011) microorganisms can impact mineral formation in several distinct ways: (a) they can modify the composition of the solution so that it becomes supersaturated or more supersaturated than it previously was with respect to a specific phase; (b) they can produce organic polymers, which can impact nucleation by favoring (or inhibiting) the stabilization of the very first mineral seeds; and (c) they can influence the impact mineral growth by inhibiting crystal growth along certain directions by production of poisoning molecules, resulting in the formation of minerals with particular shapes.

Biomineralization processes are divided into two fundamentally different groups based upon their degree of biological control (Lowenstam, 1981): (a) biologically induced mineralization (BIM), resulting from indirect modification of chemical conditions, such as a pH shift or redox transformations, in the environment by biological activity; and (b) biologically controlled mineralization (BCM), also referred to as ―organic matrix-mediated mineralization" (Lowenstam, 1981) or ―boundary-organized bio-mineralization" (Mann, 1986). BIM is an indirect mean of mineral formation and occurs as a result of metabolic activities of the organism and subsequent chemical reactions due to those metabolic activities. The mineral particles are usually formed extracellularly, have a broad particle size distribution, are poorly crystallized and do not present specific crystalline morphologies.

In BCM the organisms directs the nucleation, growth, morphology, and final location of a mineral. The mineral particles produced are characterized by being well-ordered crystals (non amorphous), having a narrow size distribution and very specific particle morphologies. These minerals are synthesized or deposited in a specific location with regard to the cell that is usually intracellular (e.g. Couradeau et al., 2012). Biologically controlled mineralization processes can be described as occurring extra-, inter- or intracellularly.

In both processes the organic components are formed first, which can then bind ions providing nucleation sites for crystal growth (Lowenstam and Weiner, 1989; Falini et al., 1996). For nucleation and growth to occur, biomineral formation requires a localized zone that achieves and maintains a sufficient supersaturation. In most biological systems, the site of mineral deposition is isolated from the environment by a physical delimiting geometry (Weiner and Dove, 2003). The term organomineralization encompasses biologically influenced and biologically induced biomineralization (Dupraz et al., 2009).

According to Dupraz et al. (2009) a third process of bacterially mediated biomineralization must be considered - the biologically influenced biomineralization, which is defined as passive mineral precipitation in the presence of organic matter, such as cell surfaces or extracellular polymeric substances (EPS), whose properties influence crystal morphology and composition. EPS consist in a complex glue-like exopolymeric matrix secreted by the cells which may vary in chemical and physical properties, but consist mainly of acidic polysaccharides with fatty acids, proteins and enzymes (Wilderer and Characklis, 1989; Costerton et al., 1995; Kawaguchi and Decho, 2002). These charged extracellular polymers may provide sites for binding of metal ions (González-Muñoz et al., 2010), and under suitable conditions may reprecipitate on the different biogenic and authigenic phases (Banerjee and Iyer, 1991).

Eukaryotes such as diatoms, foraminifers and corals have received much more attention than prokaryotes (i.e., both bacteria and archaea) in the biomineralization literature (Benzerara et al., 2011). According to the latter authors this partly results from the common assumption that eukaryote biomineralizing systems are usually determined genetically and form more complex, hence interesting, biomineral structures, while prokaryotes have a passive indirect role in mineral precipitation. However, prokaryotes, owing to the diversity of metabolic reactions they can catalyze, are able to induce the formation of minerals that are much more diverse chemically including oxides, oxi-hydroxides, carbonates, phosphates, sulfates, and sulfides.

The biomineralization mechanisms are far from fully investigated and understood for most species and aspects such as: (i) the specificity and genetic control of the diverse biomineralization mechanisms; (ii) the stability of biominerals, especially how their association with organic molecules might alter their reactivity and their transport in the environment; and (iii) biomineralization under oxidic conditions (Benzerara et al., 2011).

Mechanisms involved in biomineralization have been extensively covered by several reviews (Lowenstam, 1981; Lowenstam and Weiner, 1989; Banfield and Nealson, 1997; Bäuerlein, 2000; Mann, 2001; Skinner and Jahren, 2003; Weiner and Dove, 2003).

Nowadays the number of minerals of biological origin currently exceeds 250 (Tazaki, 2006), belonging to different mineral classes (carbonates, phosphates, sulfates, sulfides, arsenates, chlorides, fluorides, metals, oxides, hydroxides and hydrous oxides and hydrated silica), including organic crystals (by the convention of Lowenstam and Weiner, 1989). The most numerous biogenic minerals in terms of anion type are phosphates, followed by oxides, and carbonates (Tazaki, 2006).

CALCIUM CARBONATES

The calcium carbonate minerals are the most abundant biogenic minerals (Lowenstam and Weiner 1989).

Carbonate rocks represents the world's largest carbon reservoir with direct implications in the global carbon cycle and considerable carbon is sequestered in carbonate rocks forming karstic caves. Cave microorganisms are able to induce precipitation of carbonates, via biomineralization processes (Cañaveras et al., 2001; Cuezva et al., 2009) and also dissolution processes due to the excretion of acids (Northup and Lavoie, 2001; Engel, 2007). The knowledge on the importance of microorganisms in many geological processes, including speleothem formation (secondary mineral deposited in a cave by the action of water), led to an increased interest in this field over the last decade (Barton and Northup, 2007; Banks et al., 2010). Bacteria may produce carbonate minerals as a result of passive growth. Cell walls have chemically reactive sites that bind dissolved mineral-forming elements and nucleation and growth of crystals from an oversaturated solution is produced. Also mineral precipitation results from metabolic activities of bacteria. Often, bacterial activity simply triggers a change in solution chemistry that leads to oversaturation and mineral precipitation. In biological processes, oversaturation is considered as an important prerequisite for the precipitation of minerals from solution (Fortin et al., 1997), although González-Muñoz et al. (1996) estimated this is merely incidental and the critical point is the participation of cellular membranes inducing nucleation. Mineralization of calcium carbonate by bacteria isolated from caves and other subterranean environments has been reported (Groth et al., 2001; Sanchez-Moral et al., 2003a). Laiz et al. (2003) found that 61% of the *Actinobacteria* isolated from cave produced crystals in culture media. Usually, culture and field sample biominerals contain calcium carbonate polymorphs like calcite, aragonite, Mg-calcite or vaterite. Microbial species isolated from calcium carbonate minerals can produce similar crystals from organic calcium salts under laboratory conditions. In fact, Groth et al. (2001) tested the ability for producing crystals among cave bacteria, including a high number of actinobacteria. Crystal production was extensive in this group of bacteria, from which only *Gordonia* and *Nocardia* strains failed to produce crystals on B-4 medium. The tested organisms produced either calcite alone (e.g. *Brachybacterium* sp.) or both vaterite and calcite (e.g. *Rhodococcus* sp.).

Vaterite is a metastable phase that transforms to calcite or aragonite in a few hours at room temperature (Ogino et al., 1987). Vaterite is rarely found in nature but it has been identified in mortars (Gleize et al., 2000) and cultures of some microorganisms (Groth et al., 2001). Laboratory experiments have revealed that vaterite preservation is favored by the presence of organic substances, such as amino acids or mucopolysaccharides which are present in biofilms (Manoli and Dalas, 2000, Albertano and Bellezza, 2001), thus enhancing the precipitation of unstable carbonate phases (Mann et al., 1988, Pach et al., 1990, Falini et al., 1996). Vaterite preservation is also favored in phosphorus-enriched medium, because dissolved phosphorus inhibited the precipitation of calcite and/or aragonite; the stability of vaterite in the presence of phosphate ions was ascribed to the retardation of both the dissolution of vaterite and the crystallization of calcite (Katsifaras and Spanos, 1999).

In Altamira Cave, Spain, different bacterial colonizations were observed (Figure 1). Some of these adopted the form of grey round spots, isolated or in mass. According to Cuezva et al. (2012) the grey spots were able to capture atmospheric CO_2 and also bioinduce $CaCO_3$ precipitation. The microbial growth was associated with two main types of $CaCO_3$ deposits (Figure 2). Nest or rosette-like fabrics, which are aggregates of subeuhedral to euhedral calcite crystals (2-4 μm in size) with radial arrangements and delineating an external pseudohexagonal contour, and spheroidal or hemispheroidal elements averaging in diameter between 8 and 10 μm and frequently reaching diameters up to 15 μm. An attempt to isolate the bacteria involved in the grey spot formation resulted in the obtaining of strains of the genera *Streptomyces* (some of them with grey mycelia) and *Bacillus* which were found to induce $CaCO_3$ precipitation in the laboratory, as well (Laiz et al., 1999; Cañaveras et al., 1999). Similar results were obtained by Baskar et al. (2006) for bacteria isolates from stalactites in Indian caves. An example of the bacterial capacity to sequester CO_2 in solutions is carbonic anhydrase, an enzyme that catalyzes the reversible hydration of CO_2 and is widely distributed in higher organisms (animals, green algae). Several studies have linked the presence of carbonic anhydrase to bacterial uptake of atmospheric CO_2 (Smith and Ferry, 2000;

Tripp et al., 2001). Zhang et al. (2011) considered the possibility that *Bacillus mucilaginosus* could capture atmospheric CO_2 through excreted carbonic anhydrase. Thus, carbonic anhydrase may be responsible for sequestering a significant fraction of CO_2 by heterotrophic bacteria in subterranean environments.

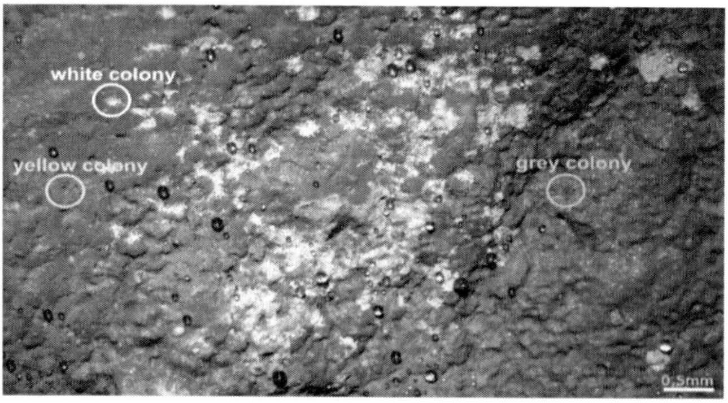

Figure 1. A wall of the Entrance Hall of Altamira Cave where the main types of bacterial colonizations: yellow-, gray and white-colored spots can be observed.

Figure 2. Scanning electron micrograph of a grey spot from Altamira Cave: detail of spheroidal elements of $CaCO_3$ above a bed of $CaCO_3$ nest-like aggregates.

In the last few years, a debate on the origin of moonmilk is attaining considerable attention. Moonmilk from carbonate speleothems has been frequently considered to be of microbial origin, either by direct precipitation of calcite by microorganisms or by the formation of a nucleation surface on which mineral precipitate (Forti, 2001; Northup and Lavoie, 2001; Cañaveras et al., 2006). A few authors have attributed the precipitation of fiber calcite crystals in moonmilk to microorganisms (Cañaveras et al. 2006; Blyth and Frisia, 2008; Curry et al., 2009; Baskar et al., 2011).

Cañaveras et al. (2006) stated that moonmilk deposits showed different microfabrics related with the different stages of moonmilk formation and suggested that bacteria were involved in this process (Figure 3). Blyth and Frisia (2008), Curry et al. (2009) and Baskar et al. (2011) reported the involvement of bacteria in moonmilk using molecular methods (Sanchez-Moral et al., 2012). However, Bindschedler et al. (2010) emphasized the fundamental role of fungi in carbonatogenesis, in spite of other authors ~~do~~ did not found chemical (Blyth and Frisia, 2008) and molecular evidences of fungal DNA (Sanchez-Moral et al., 2012) in their studies on moonmilk.

The isolation of fungi from speleothem surfaces is not an evidence of fungal involvement in a geochemical process. Recently, Vaughan et al. (2011) sampled speleothems in Kartchner Caverns, Arizona (USA), to examine fungal diversity on actively-forming speleothem surfaces. The most commonly isolated fungi belonged to the genera *Penicillium, Paecilomyces, Phialophora,* and *Aspergillus.*

Most of the genera isolated were previously found in the air from indoor environments, European cave air, and in rocks and sediments (Nováková, 2009; Porca et al., 2011). Considering the high concentration of airborne fungal spores in caves populated by bat colonies and accessible to tourism (Porca et al., 2011), most of the genera and species identified in Kartchner Caverns could correspond to airborne spores entrapped in the water layer coating the speleothems. Unfortunately, no information was provided on the metabolically active fungi in these speleothems.

Rooney et al. (2010) only isolated and identified a few fungi on moonmilk from an Irish cave: *Arthroderma melis, Aspergillus clavatus, Thysanophora penicilliodes, Cladosporium* sp. and *Penicillium kojigenum*, also common in air. Other authors have reported an absence of fungi in certain cave speleothems (Cañaveras et al., 2006; Jurado et al., 2009; Cuezva et al., 2012).

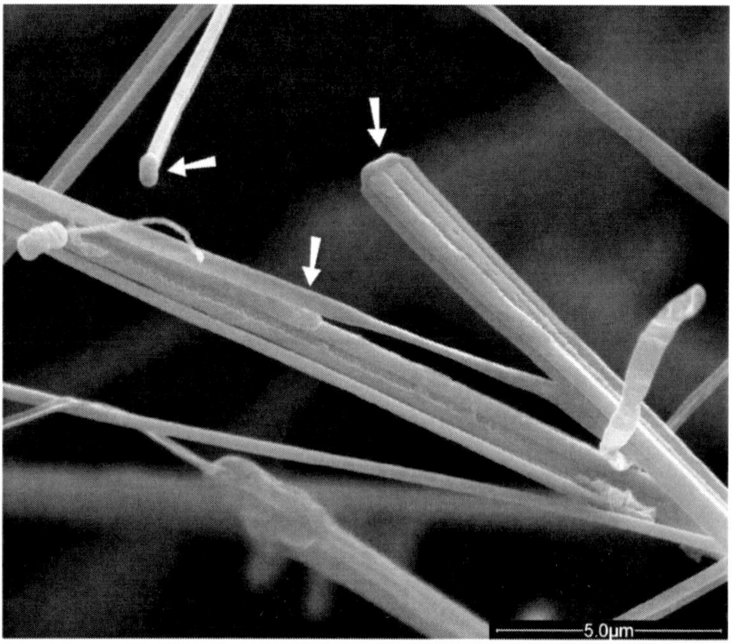

Figure 3. Scanning electron micrograph of a moonmilk deposit from Altamira cave: Genetic relationship between filaments (white arrows) and single monocrystalline rods, indicating that rods probably were formed by the direct biomineralization of bacterial filaments. Adapted from Cuezva et al. (2009).

The data presented point to a mineralization of calcium carbonate in caves, induced by bacteria, but a possible fungal involvement requires further evidence.

MANGANESE OXIDES

Manganese oxides are common minerals found in nature because of the relatively high concentration of this element in the earth's crust (Frankel and Bazylinski, 2003). The precipitation of manganese oxide and hydroxide minerals (referred hereinafter as Mn oxides) in nature is believed to be largely driven by microbiological activity, primarily bacteria and fungi (Toner et al., 2005; Webb et al., 2005a; Miyata et al., 2006). Several bacterial species are known to actively mineralize Mn oxides from including *Leptothrix discophora* (Saratovsky et al., 2006), *Metallogenium* sp. (Tebo et al., 1997), *Pseudomonas putida* (Villalobos et al., 2003), *Pedomicrobium* sp. (Ridge et al., 2007) and several marine *Bacillus* species (Webb et al., 2005a). However with the isolation of more Mn(II)-oxidizing microorganisms over the years, several research groups have found that in some surface and terrestrial environments, fungal Mn(II) oxidation may be equally or more important than bacterial oxidation (Thompson et al., 2005; Mariner et al., 2008; Saratovsky et al., 2009; Santelli et al., 2010, 2011).

The mechanism of Mn(II) oxidation and deposition is not yet understood (Webb et al., 2005b; Miyata et al., 2006; Learman et al., 2011). It is believed to be driven by an enzyme or enzyme complex involving a multicopper oxidase (Tebo et al., 2005; Miyata et al., 2006; Dick et al., 2009; Dick and Tebo, 2010), or indirectly catalyzed by bacteria by influencing chemical parameters of the microenvironment. At circumneutral pH, bacterial cell walls have negative charges due to deprotonation of the organic functional groups, which may act as nucleation sites for the aqueous metal cations (Fein, 2000; Fein et al., 2002). Recently, Learman et al. (2011) reported that bacteria can also oxidize Mn(II) through the enzymatic generation of extracellular superoxide radicals.

Biogenic Mn oxides are in general the most highly reactive and important Mn oxide phases in the environment due to their avidly capacity to adsorb trace elements (e.g. Cu, Co, Zn, Ni, Sn, Ca, Fe, Ti), including those of priority pollutants such as Pb, As and U (Nelson et al., 2002; Tebo et al., 2004, 2005; Villalobos et al., 2003, 2005a,b; Toner et al., 2006; Boonfueng et al., 2009). Nelson et al. (2002) reported that biogenic Mn oxides produced by *Leptothrix discophora* SS-1 adsorbed five times more Pb(II) than an abiotic Mn oxide, and 500 to 5000 times more than pyrolusite oxides. These findings were also inferred by Villalobos et al. (2005a,b) in a study of biogenic birnessite produced by *Pseudomonas putida* MnB1. These amply recognized sorbent characteristics of biogenic Mn oxides have stimulated interest in the development of Mn(IV) oxides for various applications in biotechnology and bioremediation (Hope and Bott, 2004; Hennebel et al., 2008; Srivastava and Majumder, 2008). The deposition of Mn oxides is a well-known occurrence in the natural environment, mainly in aquatic systems, such as hot springs, deep-sea nodules, hydrothermal sediments, fresh water sediments and streams, but less so in terrestrial environments such as the cave, karst and pseudo-karst systems (Northup et al., 2000, 2003; Spilde et al., 2005; Palmer, 2007; Frierdich et al., 2011; Rossi et al., 2011). Mn oxides in caves typically occurs by microbial processes and are often found as soft deposits forming black coatings deposited by cave stream and groundwater seepage on cave ceilings, walls and stream cobbles and pebbles (Onac et al., 1997; Northup et al., 2003;

Spilde et al., 2005; White et al., 2009; Frierdich et al., 2011). In general, the Mn-rich coatings may vary in thickness from few millimeters to many centimeters forming thick fluffy coatings, like a viscous sludge mat as described by White et al. (2009) and Miller et al. (2012a,b). These black-colored coatings are most often composed of birnessite-like minerals (Northup and Lavoie, 2001; Northup et al., 2003; Barton and Northup, 2007; White et al., 2009; Gázquez et al., 2011; Frierdich et al., 2011; Miller et al., 2012b) or todorokite (Onac et al., 1997; Northup et al., 2000, 2003; Spilde et al., 2005; Chalmin et al., 2006), which result from the activity of diverse species of Mn-oxidizing bacteria (Northup et al., 2003). Buserite has also been identified in black coatings from caves, such as in Pautler Cave, Illinois, USA (Frierdich et al., 2011). Less common but also detected in caves were lithiophorite (Spilde et al., 2005) and pyrolusite found in a Romanian cave (Onac et al., 1997) and in a 16th century granite spring water tunnel from Porto city, Portugal (Miller et al., 2012b).

The presence of Mn-oxidizing bacteria in caves can be easily evidenced by the detection of conspicuous Mn oxide deposits around or onto cell walls in samples observed under scanning electron microscopy (SEM) and/or transmission electron microscopy (TEM) combined with energy dispersive X-ray spectroscopy. For instance, crystal morphology and surface topography of samples derived from ferromanganese deposits collected in Lechuguilla and Spider Caves (New Mexico, USA) were investigated by SEM-EDS (Spilde et al., 2005). Among the five different Mn oxide morphologies described by the authors, the most prevalent was those resembling crumpled tissue paper, as well as the box-work form, which was reported to be characteristic of birnessite. Another morphology of Mn oxides was the filamentous type with 100 nm or less in diameter and hundreds of micrometers in length (Spilde et al., 2005). Miller et al. (2012b) also described nanometer Mn oxide fibrils aggregations resembling those reported by Spilde et al. (2005), and suggested to be poorly crystalline birnessite (Figure 4). Frierdich et al. (2011) described Mn oxides from seep deposits of Pautler Cave as crenulated spherical aggregates that at high magnification resembled stacked and inter-grown plate-like particles creating an intricate porous structure. In fact, all these Mn oxide features described in caves show characteristic morphologies of Mn oxides derived from natural environments (e.g. Miura and Hariya, 1997) and synthetic Mn oxide phases (e.g. Oaki and Imai, 2007; Yan et al., 2009). As reported by the former authors, the Mn oxides found in caves were usually associated with biological structures, such as bacterial filaments and EPS (Spilde et al., 2005; Miller et al., 2012a,b). It is well known that many of

Mn(II)-oxidizing bacteria precipitate Mn oxides extracellularly accumulating conspicuous oxide deposits around the cell, as is the case of the sheathed Fe-/Mn-oxidizing bacteria *Leptothrix* and budding *Pedomicrobium manganicum* bacteria (Ghiorse and Hirsch, 1979; Sly et al., 1990; Saratovsky et al., 2006). *Pedomicrobium*-like cells covered with large amounts of Mn oxides were found in ferromanganese deposit samples from a 16th century granite spring water tunnel in Porto city (Miller et al., 2012a). In addition, intriguing mineralized tubular structures rich in Mn oxides, resembling the reticulated filaments described by Melim et al. (2008) were found in this subterranean environment. The reticulated filaments had approximately 0.5 μm in diameter and up to 100 μm length, were round or flat spreading on the ferromanganese deposits and exhibited peculiar surface ornamentation resembling a honeycombed web or mesh-like fabric (Figure 5). The fine geometry of the filament walls were formed by crosslinked nanofibers with diameters of less than 100 nm, oriented parallel to the filament length and forming either individual square-shaped or diamond-shaped chambers (Figure 5). The most interesting result reported by Miller et al. (2012a) was obtained by TEM which revealed the presence of biological structures within the tubular sheaths, consisting of rod-shaped cells filling the long tubular Mn-rich sheaths. The efforts to identify this bacterium, which included clone libraries, culturing techniques and morphological comparisons failed, and the reticulated filaments could not be correlated to any known microorganism (Miller et al., 2012a), as also concluded by Melim et al. (2008) and Jones (2009).

Figure 4. FESEM image of Mn oxide aggregates in the black coatings from the 16th century granite spring water tunnel from Porto city (Portugal), embedded in EPS.

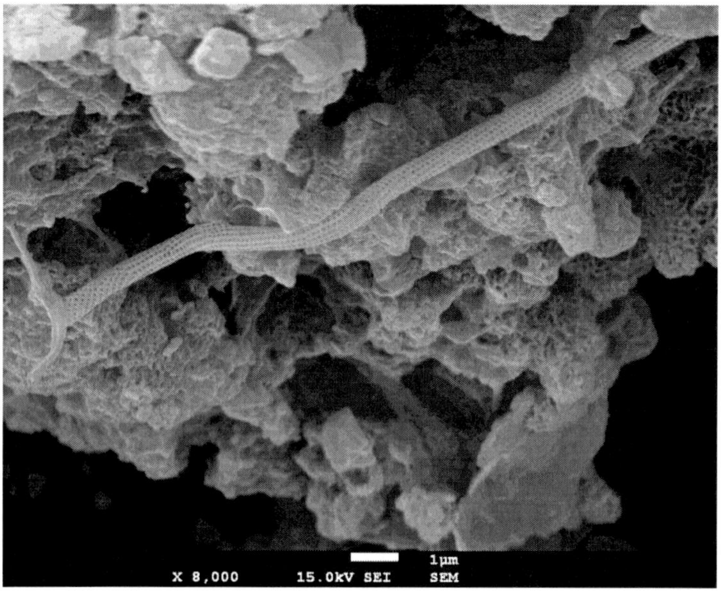

Figure 5. FESEM image depicting one Mn-rich reticulated filament within the Mn oxide aggregates found in the 16th century granite spring water tunnel from Porto city (Portugal).

Molecular biology techniques based upon 16S rRNA gene analysis have provided indirect evidence of Mn(II)-oxidizing bacteria from cave ferromanganese deposits. For instance, clone libraries from Lechuguilla or Spider Caves ferromanganese deposits showed the presence of known Mn-oxidizing bacteria, such as *Hyphomicrobium* and *Pedomicrobium manganicum* (Northup et al., 2003). Nevertheless, difficulties concerning the detection of sequences related to *Leptothrix* from environmental samples have been experienced by several authors (Takashima et al., 2008; Bruun et al., 2010; De los Ríos et al., 2011). As culturing techniques are essential for a complete assessment of the microbial communities present in an environmental sample, efforts to cultivate Mn(II)-oxidizing microorganisms from environmental samples have led to the characterization of Mn oxidation mechanisms and the discovery of bacteria with distinct traits (Sly et al., 1990; Toner et al., 2005; Webb et al., 2005b; Miyata et al., 2006; Johnson and Tebo, 2008; Learman et al., 2011). Isolation of putative Mn(II)-oxidizing microorganisms from caves, followed by production of Mn oxides in the laboratory, is necessary to understand the mechanisms involved in microbial oxidation.

IRON OXIHYDROXIDES

In the 1830s, iron (Fe) bacteria were among the first groups of microbes to be recognized for carrying out a fundamental geological process, namely the oxidation of iron. Nowadays it is amply recognized that the oxidation of Fe(II) to insoluble Fe(III) oxihydroxides, referred to hereinafter as Fe oxides, can be carried out by phyllogenetically diverse microorganisms (Emerson et al., 2010), but in caves is particularly associated with *Gallionella ferruginea* (Hallberg and Ferris, 2004) and the Fe-/Mn-oxidizing *Leptothrix* species (Emerson et al., 2010). Caldwell and Caldwell (1980) reported Fe-rich filaments with hexagonal matrix onto the bacterial sheaths collected from an iron-containing pool in Carlsbad Cavern (New Mexico, USA), resembling the Mn/Fe-oxidizing bacteria *Leptothrix*. De los Ríos et al. (2011) reported that *Gallionella* and *Leptothrix* were the most frequently observed morphotypes and probably the main contributors to speleothem formation in lava tubes from Terceira Island (Portugal). Under oxygenic conditions, these bacteria can oxidize Fe(II) to Fe(III) and precipitate the biogenic Fe oxides extracellularly near or on the bacterial cells (Fortin and Langley, 2005), helping to form features such as mineral deposits and speleothems. Fe oxides can be found in subsurface environments as wall, ceiling and floor coatings/crusts and sometimes as their own speleothems, exhibiting yellowish, orange, reddish, ochreous and brown color (Figure 6). Usually they are soft, friable and porous stalactites and stalagmites (Baskar et al., 2008; De los Ríos et al., 2011). Moreover, the Fe oxides can be found associated with Mn oxides forming multicolored ferromanganese deposits (Northup et al., 2003; Spilde et al., 2005). White et al. (2009) demonstrated that in 39 dark deposits collected from several caves in USA at least some Fe was present, stating that it is not appropriate to label every black deposits seen in caves as Mn oxides.

As in the case of Mn oxides, Fe oxides are listed among cave minerals of predominantly microbial origin (Spilde et al., 2005; De los Ríos et al., 2011). For instance and as mentioned before, the ferromanganese deposits found in Lechuguilla Cave (Northup et al., 2003) were hypothesized to have formed from microorganisms extracting Mn and Fe from the limestone bedrock and concentrating it at the rock surface (Spilde et al., 2005). Several iron oxides were found in the ferromanganese deposits with distinctive morphologies, such as star-shaped which were identified as goethite, tabular to blocky hexagonal crystals of hematite, bottlebrush-like morphology and others.

Figure 6. Orange/reddish Fe-rich speleothems developing on the granite ceiling and walls of the 16th century spring water tunnel in Porto city, Portugal.

Baskar et al. (2008) studying the microbial mats and their possible influence on iron mineral precipitation found in the spring waters of the Borra caves, Vishakapatanam (India), corroborated that the iron mineral deposits were based on microbe–mineral interactions rather than a pure inorganic precipitation reaction. The SEM examinations of the mats showed the occurrence of filamentous bacteria, being the dominant species sheathed Fe-/Mn-oxidizing bacteria (Baskar et al., 2008). Frierdich et al. (2011) investigated the occurrence, composition, and structure of nanocrystalline Fe and Mn oxide minerals in the Pautler Cave and explored the role of these mineral phases in controlling trace element and heavy metal mobility. The authors showed that one of the studied sites predominantly contained the typical twisted *Gallionella* filaments and masses of ferrihydrite. Numerous filaments with elongated tubular forms covered with ferrihydrite, probably of *Leptothrix*, were observed in another site of the Pautler Cave, which led the researchers to corroborate that microorganisms were responsible to the deposition of the iron oxide deposits. In addition, it was demonstrated that these biogenic oxides, together with Mn oxides, scavenged heavy metals and trace elements from groundwater (Frierdich et al., 2011).

Fe and Mn oxides formed in caves often exhibit the ability to concentrate heavy metals and potentially toxic trace elements by several orders of magnitude (Onac et al., 1997). Caves in Colorado, South Dakota, and Arizona were reported to contain Fe oxides with As and Pb concentrations as high as 15000 and 8000 ppm, respectively (Luiszer, 1995). Consequently, the highly reactivity of Fe and Mn oxide minerals can control the distribution and speciation of trace elements in subsurface rock environments.

OTHER MINERALS

Numerous biogenic minerals such as phosphates and sulfates have been reported in caves and subterranean environments such as witherite (barium carbonate), struvite (magnesium ammonium phosphate), hydromagnesite (hydrated magnesium carbonate) or brushite (dicalcium phosphate dihydrate). In fact, struvite is formed by *Agromyces ramosus*, *Bacillus* sp. and *Ralstonia metallidurans*, bacteria which were isolated from Roman catacombs (Sanchez-Moral et al., 2003b). The finding of struvite in cultures of catacomb isolates and the presence of appreciable amounts of phosphorus and magnesium in the tuff led Sanchez-Moral et al. (2003b) to consider that some phosphate minerals found in the catacombs could have a biological origin. Recently, Laiz et al. (2009) observed the precipitation of struvite in cultures of *Rubrobacter* spp. isolated from a Roman necropolis.

Microbial communities, where *Streptomyces* species predominate, were found in association with hydromagnesite and needle-fiber aragonite deposits in Altamira Cave (Cañaveras et al., 1999). Both hydromagnesite crystals and aragonite needle-fibers were usually embedded in mucous or filamentous microbial biofilms. In the moonmilk deposits the hydromagnesite crystals were usually platy or blade-like in shape and arranged as spherular or rosette-like clusters (resembling miniature desert roses) or as house of cards aggregates.

Extensive brushite moonmilk flowstone has been reported from Kartchner Caverns (Hill, 1999). This derives from decaying bat guano in an acid-rich environment (Hill and Forti, 1997). It is remarkable the finding of brushite in cultures of *Bacillus* sp. and *Ralstonia metallidurans*, which previously has not been identified so far in bacterial culture media.

Witherite, a naturally occurring barium carbonate was produced by *Agromyces ramosus* (Sanchez-Moral et al., 2004). It has been reported that Ba and Sr can replace Ca in carbonates, and apparently bacteria can produce Ba and Sr carbonates when these elements are provided as organic salts (e.g. acetate).

Conclusion

The geomicrobiology of subsurface rock environments is far from fully investigated and understood even though it may be of great interest and could provide useful information about general and applied microbiology and geosciences. Analysis of microbial communities, their role in bio-mineralization processes as well as their geomicrobiological interactions can help to understand the steps of colonization, the subsurface microbial diversity, the biominerals diversity and the various biomineralization mechanisms. The data presented in this chapter indicate a geomicrobiological contribution in the precipitation of Ca, Mn, Fe, Ba, Sr and Mg-carbonates in caves and subterranean environments. Likely, this could be extended to some other elements that would be specific of a definite subterranean environment.

Acknowledgments

This paper was partially financed by FEDER Funds through *Programa Operacional Factores de Competitividade* - COMPETE and by Portuguese Funds through FCT- *Fundação para a Ciência e a Tecnologia* (PEst-OE/CTE/UI0098/2011) and by Spanish Ministry of Economy and Competitiveness (Project CGL2011-2569). AZM would like to thank the FCT for her postdoctoral fellowship (SFRH/BPD/63836/2009). This work was also supported by Consolider project TCP CSD2007-00058.

References

Albertano, P., Bellezza, S., 2001. Cytochemistry of cyanobacterial exopolymers in biofilms from Roman hypogea. *Nova Hedwigia* 123, 501-518.

Anderson, C.R., James, R.E., Fru, E.C., Kennedy, C.B., Pedersen, K., 2006. *In situ* ecological development of a bacteriogenic iron oxide-producing microbial community from a subsurface granitic rock environment. *Geobiology* 4, 29-42.

Banerjee, R., Iyer, S.D., 1991. Biogenic influence on the growth of ferromanganese micronodules in the Central Indian Basin. *Mar. Geol.* 97, 413-421.

Banfield, J.F., Nealson, K.H., 1997. Geomicrobiology. In: Reviews in Mineralogy, Geological Society of America, Boulder, CO, USA, v. 35, p. 448.

Banks, E.D., Taylor, N.M., Gulley, J., Lubbers, B.R., Giarrizzo, J.G., Bullen, H.A., Hoehler, T.M., Barton, H.A., 2010. Bacterial calcium carbonate precipitation in cave environments: a function of calcium homeostasis. *Geomicrobiol. J.* 27, 444-454.

Barton, H.A., Spear, J.R., Pace, N.R., 2001. Microbial life in the underworld: Biogenicity in secondary mineral formations. *Geomicrobiol. J.* 18, 359-368.

Barton, H.A., 2006. Introduction to cave microbiology: a review for the non-specialist. *Journal of Cave and Karst Studies* 68, 43-54.

Barton, H.A., Northup, D.E., 2007. Geomicrobiology in cave environments: past, current and future perspectives. *J. Cave Karst Stud.* 69, 163-178.

Baskar, S., Baskar, R., Mauclaire, L., McKenzie, J.A., 2006. Microbially induced calcite precipitation in culture experiments: Possible origin for stalactites in Sahastradhara caves, Dehradun, India. *Curr. Sci. India* 90, 58-64.

Baskar, S., Baskar, R., Lee, N., Kaushik, A., Theophilus, P.K., 2008. Precipitation of iron in microbial mats of the spring waters of Borra caves: some geomicrobiological aspects. *Environ. Geol.* 56, 237-243.

Baskar, S., Baskar, R., Routh, J., 2011. Biogenic evidences of moonmilk deposition in the Mawmluh Cave, Meghalaya, India. *Geomicrobiol. J.* 28, 252-265.

Bäuerlein, E., 2000. Biomineralization: From Biology to Biotechnology and Medical Application, Wiley-VHC Verlag GmbH & Co. KgaA, Weinheim, Germany.

Bäuerlein, E., 2004. Biomineralization: Progress in Biology, Molecular Biology and Application, Wiley-VHC Verlag, Weinheim, Germany.

Benzerara, K., Miot, J., Morin, G., Ona-Nguema, G., Skouri-Panet, F., Ferard, C., 2011. Significance, mechanisms and environmental implications of microbial biomineralization. C. R. Geosci. 343, 160-167.

Blyth, A.J., Frisia, S., 2008. Molecular evidence for bacterial mediation of calcite formation in cold high-altitude caves. *Geomicrobiol. J.* 25, 101-111.

Boonfueng, T., Axe, L., Yee, N., Hahn, D., Ndiba, P.K., 2009. Zn sorption mechanisms onto sheathed *Leptothrix discophora* and the impact of the nanoparticulate biogenic Mn oxide coating. *J. Colloid Interf. Sci.* 333, 439-447.

Bruun, A-M., Finster, K., Gunnlaugsson, H.P., Nornberg, P., Friedrich, M.W., 2010. A comprehensive investigation on iron cycling in a freshwater seep including microscopy, cultivation and molecular community analysis. *Geomicrobiol. J.* 27, 15-34.

Caldwell, D.E., Caldwell, S.J., 1980. Fine structures of in situ microbial iron deposits. *Geomicrobiol. J.* 2, 39-53.

Cañaveras, J.C., Hoyos, M., Sanchez-Moral, S., Sanz-Rubio, E., Bedoya, J., Soler, V., Groth, I., Schumann, P., Laiz, L., Gonzalez, I., Saiz-Jimenez, C., 1999. Microbial communities associated with hydromagnesite and needle-fiber aragonite deposits in a karstic cave (Altamira, Northern Spain). *Geomicrobiol. J.* 16, 9-25.

Cañaveras, J.C., Sanchez-Moral, S., Soler, V., Saiz-Jimenez, C., 2001. Microorganisms and microbially induced fabrics in cave walls. *Geomicrobiol. J.* 18, 223-240.

Cañaveras, J.C., Cuezva, S., Sanchez-Moral, S., Lario, J., Laiz, L., Gonzalez, J.M., Saiz-Jimenez, C., 2006. On the origin of fiber calcite crystals in moonmilk deposits. *Naturwissenschaften* 93, 27-32.

Chalmin, E., Vignaud, C., Salomon, H., Farges, F., Susini, J., Menu, M., 2006. Minerals discovered in paleolithic black pigments by transmission electron microscopy and micro-X-ray absorption near-edge structure. *Appl. Phys. A–Mater.* 83, 213-218.

Costerton, J.W., Lewandowski, Z., Caldwell, D.E., Korber, D.R., Lappin-Scott, H.M., 1995. Microbial biofilms. *Annu. Rev. Microbiol.* 49, 711-745.

Couradeau, E., Benzerara, K., Gérard, E., Moreira, D., Bernard, S., Brown jr., G.E., López-García, P., 2012. An early-branching microbialite cyanobacterium forms intracellular carbonates. *Science* 336, 459-462.

Cuezva, S., Sanchez-Moral, S., Saiz-Jimenez, C., Cañaveras, J.C., 2009. Microbial communities and associated mineral fabrics in Altamira Cave, *Spain. Int. J. Speleol.* 38, 83-92.

Cuezva, S., Fernandez-Cortes, A, Porca, E., Pašić, L., Jurado, V., Hernandez-Marine, M., Serrano-Ortiz, P., Hermosin, B., Cañaveras, J.C., Sanchez-Moral, S., Saiz-Jimenez, C., 2012. The biogeochemical role of

Actinobacteria in Altamira Cave, Spain. *FEMS Microbiol. Ecol.* 81, 281-290.

Curry, M.D., Boston, P.J., Spilde, M.N., Baichtal, J.F., Campbell, A.R., 2009. Cottonballs, a unique subaqueous moonmilk, and abundant subaerial moonmilk in Cataract Cave, Tongass National Forest, Alaska. I. *J. Speleol.* 38, 111-128.

De los Ríos, A., Bustillo, M.A., Ascaso, C., Carvalho, M.R., 2011. Bioconstructions in ochreous speleothems from lava tubes on Terceira Island (Azores). *Sediment. Geol.* 236, 117-128.

Dick, G.J., Clement, B.G., Webb, S.M., Fodrie, F.J., Bargar, J.R., Tebo, B.M., 2009. Enzymatic microbial Mn(II) oxidation and Mn biooxide production in the Guaymas Basin deep-sea hydrothermal plume. *Geochim. Cosmochim. Ac.* 73, 6517-6530.

Dick, G.J., Tebo, B.M., 2010. Microbial diversity and biogeochemistry of the Guaymas Basin deep-sea hydrothermal plume. *Environ. Microbiol.* 12, 1334-1347.

Dupraz, C., Reid, R.P., Braissant, O., Decho, A.W., Norman, R.S., Visscher, P.T., 2009. Processes of carbonate precipitation in modern microbial mats. *Earth Sci. Rev.* 96, 141-162.

Emerson, D., Fleming, E.J., McBeth, J.M., 2010. Iron-oxidizing bacteria: an environmental and genomic perspective. *Annu. Rev. Microbiol.* 64, 561-583.

Engel, A.S., 2007. Observations on the biodiversity of sulfidic karst habitats. *J. Cave Karst Stud.* 69,187-206.

Falini, G., Albeck, S., Weiner, S., Addadi, L., 1996. Control of aragonite or calcite polymorphism by mollusk shell macromolecules. *Science* 271, 67-69.

Fein, J.B., 2000. Quantifying the effects of bacteria on adsorption reactions in water-rock systems. *Chem. Geol.* 169, 265-280.

Fein, J.B., Scott, S., Rivera, N., 2002. The effect of Fe on Si adsorption by *Bacillus subtilis* cell walls: insights into non-metabolic bacterial precipitation of silicate minerals. *Chem. Geol.* 182, 265-273.

Forti, P., 2001. Biogenic speleothems: an overwiew. *Int. J. Speleol.* 30, 39-56.

Fortin, D., Ferris, F.G., Beveridge, T.J., 1997. Surface-mediated mineral development by bacteria. In: Banfield, J.F., Nealson, K.H. (Eds) Geomicrobiology: Interactions between Microbes and Minerals. Mineralogy Society of America, Washington, pp. 161-180.

Fortin, D., Langley, S., 2005. Formation and occurrence of biogenic iron-rich minerals. *Earth-Sci. Rev.* 72, 1-19.

Foster, A.L., Ashley, R.P., 2002. Characterization of arsenic species in microbial mats from an inactive gold mine. *Geochem. Explor. Env.* A 2, 253-261.

Frankel, R.B., Bazylinski, D.A., 2003. Biologically induced mineralization by bacteria. *Rev. Mineral. Geochem.* 54, 95-114.

Frierdich, A.J., Hasenmueller, E.A., Catalano, J.G., 2011. Composition and structure of nanocrystalline Fe and Mn oxide cave deposits: Implications for trace element mobility in karst systems. *Chem. Geol.* 284, 82-96.

Gázquez, F., Calaforra, J.M., Forti, P., 2011. Black Mn-Fe crusts as markers of abrupt palaeoenvironmental changes in El Soplao Cave (Cantabria, Spain). *Int. J. Speleol.* 40, 163-169.

Ghiorse, W.C., Hirsch, P., 1979. An ultrastructural study of iron and manganese deposition associated with extracellular polymers of *Pedomicrobium*-like budding bacteria. *Arch. Microbiol.* 123, 213-226.

Gleize, P., Silva, D.A., Nappi, S., 2000. Ancient rendering mortars from a Brazilian place. Its characteristics and microstructure. *Cement Concrete Res.* 30, 1609-1614.

González-Muñoz, M.T., Rodriguez-Navarro, C., Martínez-Ruiz, F., Arias, J.M., Merroun, M.L., Rodriguez-Gallego, M., 2010. Bacterial biomineralization: new insights from *Myxococcus*-induced mineral precipitation. In: Pedley, H.M., Rogerson, M. (Eds.) Tufas and Speleothems: Unravelling the Microbial and Physical Controls. Geological Society, London, Special Publications, v. 336, pp. 31-50.

Groth, I., Schumann, P., Laiz, L., Sanchez-Moral, S., Cañaveras, J.C., Saiz-Jimenez, C., 2001. Geomicrobiological study of the Grotta dei Cervi, Porto Badisco, Italy. *Geomicrobiol. J.* 18, 241-258.

Hallberg, R., Ferris, F.G., 2004. Biomineralization by *Gallionella*. *Geomicrobiol. J.* 21, 325-330.

Hennebel, T., de Gusseme, B., Boon, N., Verstraete, W., 2008. Biogenic metals in advanced water treatment. *Trends Biotechnol.* 27, 90-98.

Hill, C.A., Forti, P., 1997. Cave Minerals of the World, 2nd ed. National Speleological Society, Huntsville, Alabama.

Hill, C.A., 1999. Mineralogy of Kartchner Caverns, Arizona. *J. Cave Karst Stud.* 61, 73-78.

Hofmann, B.A., Farmer, J.D., von Blanckenburg, F., Fallick, A.E., 2008. Subsurface filamentous fabrics: an evolution of origins based on morphological and geochemical criteria, with implications for exopaleontology. *Astrobiology* 8, 87-117.

Hope, C.K., Bott, T.R., 2004. Laboratory modelling of manganese biofiltration using biofilms of *Leptothrix discophora*. *Water Res.* 38, 1853-1861.

Jimenez-Lopez, C., Jroundi, F., Rodríguez-Gallego, M., Arias, J.M., Gonzalez-Muñoz, M.T., 2007. Biomineralization induced by *Myxobacteria*. In: Méndez-Vilas, A. (Ed.) Communicating current research and educational topics and trends in applied microbiology. Formatex, Badajoz, v.1, pp. 143-154.

Johnson, H.A., Tebo, B.M., 2008. In vitro studies indicate a quinone is involved in bacterial Mn(II) oxidation. *Arch. Microbiol.* 189, 59-69.

Jones, B., 2009. Cave pearls - The integrated product of abiogenic and biogenic processes. *J. Sediment. Res.* 79, 689-710.

Jurado, V., Fernandez-Cortes, A., Cuezva, S., Laiz, L., Cañaveras, J.C., Sanchez-Moral, S., Saiz-Jimenez, C., 2009. The fungal colonization of rock art caves. *Naturwissenschaften* 96, 1027-1034.

Katsifaras, A., Spanos, N., 1999. Effect of inorganic phosphate ions on the spontaneous precipitation of vaterite and on the transformation of vaterite to calcite. *J. Cryst. Growth* 204, 183-190.

Kawaguchi, T., Decho, A.W., 2002. In situ analysis of carboxyl (-COOH) and sulfhydral (-SH) groups of extracellular polymeric secretions (EPS) by confocal scanning laser microscopy. *Anal. Biochem.* 304, 266-267.

Laiz, L., Groth, I., Gonzalez, I., Saiz-Jimenez, C., 1999. Microbiological study of the dripping waters in Altamira cave (Santillana del Mar, Spain). *J. Microbiol. Meth.* 36, 129-138.

Laiz, L., Gonzalez, J.M., Saiz-Jimenez, C., 2003. Microbial communities in caves: Ecology, physiology, and effects on paleolithic paintings. In: Koestler, R.J., Koestler, V.R., Carola, A.E., Nieto-Fernández, F.E. (Eds) Art, Biology, and Conservation: Biodeterioration of works of Art. The Metropolitan Museum of Art, New York, pp. 210-225.

Laiz, L., Miller, A.Z., Gonzalez, J.M., Akatova, E., Jurado, V., Dionísio, A., Macedo, M.F., Saiz-Jimenez, C., 2009. Isolation of five *Rubrobacter* strains from biodeteriorated monuments. *Naturwissenschaften* 96, 71-79.

Learman, D.R., Voelker, B.M., Vazquez-Rodriguez, A.I., Hansel, C.M., 2011. Formation of manganese oxides by bacterially generated superoxide. *Nat. Geosci.* 4, 95-98.

Lowenstam, H.A., 1981. Minerals formed by organisms. *Science* 211, 1126-1131.

Lowenstam, H.A., Weiner, S., 1989. On Biomineralization. Oxford University Press, New York.

Luiszer, F.G., 1995. Study of iron and manganese oxide deposits from Jewel and Jasper Caves, South Dakota, Cave of the Winds, Ocher Cave, Fairy Cave, Premonition Cave and Porcupine Cave, Colorado, and two caves in Arizona. *Nat. Spel. Soc. Bull.* 57, 72.

Mann, S., 1986. On the nature of boundary-organised biomineralization. *J. Inorg. Chem.* 28, 363-371.

Mann, S., Heywood, B.R., Rajam, S., Birchall, J.D. 1988. Controlled crystallization of $CaCO_3$ under stearic acid monolayers. *Nature* 334, 692-695.

Mann, S., 2001. Biomineralization: principles and concepts in bioinorganic materials chemistry. Oxford University Press, New York.

Manoli, F., Dalas, E., 2000. Spontaneous precipitation of calcium carbonate in the presence of chondroitin sulfate. *J. Cryst. Growth* 217, 416-421.

Mariner, R., Johnson, D., Hallberg, K., 2008. Characterisation of an attenuation system for the remediation of Mn (II) contaminated waters. *Hydrometallurgy* 94, 100-104.

Melim, L.A., Shinglman, K.M., Boston, P.J., Northup, D.E., Spilde, M.N., Queen, J.M., 2001. Evidence for microbial involvement in pool finger precipitation, Hidden Cave, New Mexico: *Geomicrobiol. J.* 18, 311-329.

Melim, L.A., Northup, D.E., Spilde, M.N., Jones, B., Boston, P.J. Bixby, R.J., 2008. Reticulated filaments in cave pool speleothems: microbe or mineral? *J. Cave Karst Stud.* 70, 135-141.

Miller, A.Z., Hernández-Mariné, M., Jurado, V., Dionísio, A., Barquinha, P., Fortunato, E., Afonso, M.J., Chaminé, H.I., Saiz-Jimenez, C., 2012a. Enigmatic reticulated filaments in subsurface granite. *Environ. Microbiol. Rep.*, doi:10.1111/j.1758-2229.2012.00375.x.

Miller, A.Z., Dionísio, A., Sequeira Braga, M.A., Hernández-Mariné, M., Afonso, M.J., Muralha, V.S.F., Herrera, L.K., Raabe, J., Fernandez-Cortes, A., Cuezva, S., Hermosin, B., Sanchez-Moral, S., Chaminé, H., Saiz-Jimenez, C., 2012b. Biogenic Mn oxide minerals coating in a subsurface granite environment. *Chem. Geol.*, 322-323, 181-191.

Miura, H., Hariya, Y., 1997. Recent manganese oxide deposits in Hokkaido, Japan. Geological Society, London, Special Publications, 119, pp. 281-299.

Miyata, N., Tani, Y., Maruo, K., Tsuno, H., Sakata, M., Iwahori, K., 2006. Manganese(IV) oxide production by *Acremonium* sp. strain KR21–2 and extracellular Mn(II) oxidase activity. *Appl. Environ. Microbiol.* 72, 6467-6473.

Nealson, K.H., Stahl, D.A., 1997. Microorganisms and biogeochemical cycles; what can we learn from layered microbial communities? In: Banfield, J.F., Nealson, K.H., (Eds.) Geomicrobiology: Interactions between microbes and minerals. Mineralogical Society of America, Washington, DC, v. 35, pp. 5-34.

Nelson, Y.M., Lion, L.W., Shuler, M.L., Ghiorse, W.C., 2002. Effect of oxide formation mechanisms on lead adsorption by biogenic manganese (hydr)oxides, iron (hydr)oxides, and their mixtures. *Environ. Sci. Technol.* 36, 421-425.

Newman, D.K., Banfield, J.F., 2002. Geomicrobiology: how molecular-scale interactions underpin biogeochemical systems. *Science* 296, 1071-1077.

Northup, D.E., Dahm, C.N., Melim, L.A., Spilde, M.N., Crossey, L.J., Lavoie, K.H., Mallory, L.M., Boston, P.J., Cunningham, K.I., Barns, S.M., 2000. Evidence for geomicrobiological interactions in Guadalupe caves. *J. Cave Karst Stud.* 62, 80-90.

Northup, D.E., Lavoie, K.H., 2001. Geomicrobiology of caves: a review. *Geomicrobiol. J.* 18, 199-222.

Northup, D.E., Barns, S.M., Yu, L.E., Spilde, M.N., Schelble, R.T., Dano, K.E., Crossey, L.J., Connolly, C.A., Boston, P.J., Natvig, D.O., Dahm, C.N., 2003. Diverse microbial communities inhabiting ferromanganese deposits in Lechuguilla and Spider Caves. *Environ. Microbiol.* 5, 1071-1086.

Nováková, A., 2009. Microscopic fungi isolated from the Domica Cave System (Slovak Karst National Park, Slovakia). A review. *Int. J. Speleol.* 38, 71-82.

Oaki Y., Imai, H., 2007. One-pot synthesis of manganese oxide nanosheets in aqueous solution: chelation-mediated parallel control of reaction and morphology. *Angew. Chem. Int. Ed.* 46, 4951-4955.

Ogino, T., Suzuki, T., Sawada, R., 1987. The formation and transformation mechanism of calcium carbonate in water. *Geochim. Cosmochim. Acta* 51, 2757-2767.

Onac, B.P., Pedersen, R.B., Tysseland, M., 1997. Presence of rare-earth elements in black ferromanganese coatings from Vântului Cave. *J. Cave Karst Stud.* 59, 128-131.

Pach, L., Hrabe, Z., Komarneni, S., Roy, R., 1990. Controlled crystallization of vaterite from viscous solutions of organic colloids. *J. Mater. Res.* 5, 2928-2932.

Palmer, A.N., 2007. Cave Geology. Cave Books, Dayton.

Pedersen, K., Arlinger, J., Ekendahl, S., Hallbeck, L., 1996. 16S rRNA gene diversity of attached and unattached bacteria in boreholes along the access tunnel to the Äspo hard rock laboratory, Sweden. *FEMS Microbiol. Ecol.* 19, 249-262.

Porca, E., Jurado, V., Martín-Sánchez, P.M., Hermosin, B., Bastián, F., Alabouvette, C., Saiz-Jimenez, C., 2011. Aerobiology: an ecological indicator for early detection and control of fungal outbreaks in caves. *Ecol. Indic.* 11, 1594-1598.

Ridge, J.P., Lin, M., Larsen, E.I., Fegan, M., McEwan, A.G., Sly, L.I., 2007. A multicopper oxidase is essential for manganese oxidation and laccase-like activity in *Pedomicrobium* sp. ACM 3067. *Environ. Microbiol.* 9, 944-953.

Rooney, D.C., Hutchens, E., Clipson, N., Baldini, J., McDermott, F., 2010. Microbial community diversity of moonmilk deposits at Ballynamintra Cave, Co. Waterford, Ireland. *Microb. Ecol.* 60, 753-61.

Rossi, C., Lozano, R.P., Isanta, N., Hellstrom, J., 2011. Manganese stromatolites in caves: El Soplao (Cantabria, Spain). *Geology* 38, 1119-1122.

Sanchez-Moral S., Cañaveras J.C., Laiz L., Saiz-Jimenez C., Bedoya J., Luque, L., 2003a. Biomediated precipitation of calcium carbonate metastable phases in hypogean environments: a short review. *Geomicrobiology* J. 20, 491-500.

Sanchez-Moral, S., Bedoya, J., Luque, L., Cañaveras, J.C., Jurado, V., Laiz, L., SaizJimenez, C., 2003b. Biomineralization of different crystalline phases by bacteria isolated from catacombs. In: Saiz-Jimenez C. (Ed.) Molecular Biology and Cultural Heritage, Balkema, Lisse, pp. 179-185.

Sanchez-Moral, S., Luque, L., Cañaveras, J.C., Laiz, L., Jurado, V., Hermosin, B., Saiz-Jimenez, C., 2004. Bioinduced barium precipitation in St. Callixtus and Domitilla catacombs. *Ann. Microbiol.* 54, 1-12.

Sanchez-Moral, S., Portillo, M.C., Janices, I., Cuezva, S., Fernández-Cortés, A., Cañaveras, J.C., Gonzalez, J.M., 2012. The role of microorganisms in the formation of calcitic moonmilk deposits and speleothems in Altamira Cave. *Geomorphol.* 139-140, 285-292.

Santelli, C.M., Pfister, D., Lazarus, D., Sun, L., Burgos, W.D., Hansel, C.M., 2010. Promotion of Mn(II) oxidation and remediation of coal mine drainage in passive treatment systems by diverse fungal and bacterial communities. *Appl. Environ. Microbiol.* 76, 4871–4875.

Santelli, C.M., Webb, S.M., Dohnalkova, A.C., Hansel, C.M., 2011. Diversity of Mn oxides produced by Mn(II)-oxidizing fungi. *Geochim. Cosmochim. Ac.* 75, 2762-2776.

Saratovsky, I., Wightman, P.G., Pasten, P.A., Gaillard, J.F., Poeppelmeier, K.R., 2006. Manganese oxides: parallels between abiotic and biotic structures. *J. Am. Chem. Soc.* 128, 11188-11198.

Saratovsky, I., Gurr, S., Hayward, M., 2009. The structure of manganese oxide formed by the fungus *Acremonium* sp. Strain KR21–2. *Geochim. Cosmochim. Ac.* 73, 3291-3300.

Schabereiter-Gurtner, C., Saiz-Jimenez, C., Piñar, G., Lubitz, W., Rölleke, S., 2004. Phylogenetic diversity of bacteria associated with Paleolithic paintings and surrounding rock walls in two Spanish caves (Llonin and Garma). *FEMS Microbiol. Ecol.* 47, 235-247.

Skinner, H.C.W., Jahren, A.H., 2003. Biomineralization. *Treatise on Geochemistry* 8, 117-184.

Sly, L.I., Arunpairojana, V., Dixon, D.R., 1990. Binding of colloidal MnO_2 by extracellular polysaccharides of *Pedomicrobium manganicum. Appl. Environ. Microbiol.* 56, 2791-2794.

Smith, K.S., Ferry, J.G., 2000. Prokaryotic carbonic anhydrases. FEMS *Microbiol. Rev.* 24, 335-366.

Spilde, M.N., Northup, D., Boston, P.J., Schelble, R.T., Dano, K.E., Crossey, L.J., Dahm, C.N., 2005. Geomicrobiology of cave ferromanganese deposits: a field and laboratory investigation. *Geomicrobiol. J.* 22, 99-116.

Srivastava, N.K., Majumder, C.B., 2008. Novel biofiltration methods for the treatment of heavy metals from industrial wastewater. *J. Hazard. Mater.* 151, 1-8.

Takashima, C., Kano, A., Naganuma, T., Tazaki, K., 2008. Laminated iron texture by iron-oxidizing bacteria in a calcite travertine. *Geomicrobiol. J.* 25, 193-202.

Tazaki, K., 2006. Clays, microorganisms, and biomineralization. In: Bergaya, F., Theng, B.K.G., Lagaly, G. (Eds.) Handbook of clay science. Elsevier, Amsterdam, pp. 477-497.

Tebo, B.M., Ghiorse, W.C., van Waasbergen, L.G., Siering, P.L., Caspi, R., 1997. Bacterially mediated mineral formation: Insights into manganese (II) oxidation from molecular genetic and biochemical studies. *Rev. Mineral.* 35, 225-266.

Tebo, B.M., Bargar, J.R., Clement, B.G., Dick, D.J., Murray, K.J., Parker, D., Verity, R., Webb, S.M., 2004. Biogenic manganese oxides: properties and mechanisms of formation. *Annu. Rev. Earth Pl. Sc.* 32, 287–328.

Tebo, B.M., Johnson, H.A., McCarthy, J.K., Templeton, A.S., 2005. Geomicrobiology of manganese(II) oxidation. *Trends Microbiol.* 13, 421-428.

Thompson, I., Huber, D., Guest, C., Schulze, D., 2005. Fungal manganese oxidation in a reduced soil. *Environ. Microbiol.* 7, 1480-1487.

Toner, B., Fakra, S., Villalobos, M., Warwick, T., Sposito, G., 2005. Spatially resolved characterization of biogenic manganese oxide production with a bacterial biofilm. *Appl. Environ. Microbiol.* 71, 1300-1310.

Toner, B., Manceau, A., Webb, S.M., Sposito, G., 2006. Zinc sorption to biogenic hexagonal-birnessite particles within a hydrated bacterial biofilm. *Geochim. Cosmochim.* Ac. 70, 27-43.

Tripp, B.C., Smith, K., Ferry, J.G., 2001. Carbonic anhydrase: new insights for an ancient enzyme. *J. Biol. Chem.* 276, 48615-48618.

Van Cappellen, P., 2003. Biomineralization and global biogeochemical cycles. In: Dove, P.M., Weiner, S., De Yoreo, J.J. (Eds.) Biomineralization. Mineralogical Society of America, Washington, DC, v. 54, pp. 357-381.

Vaughan, M.J., Maier, R.M., Pryor, B.M., 2011. Fungal communities on speleothem surfaces in Kartchner Caverns, Arizona, USA. *Int. J. Speleol.* 40, 65-77.

Villalobos, M., Toner, B., Bargar, J., Sposito, G., 2003. Characterization of the manganese oxide produced by *Pseudomonas putida* strain MnB1. *Geochim. Cosmochim. Ac*. 67, 2649-2662.

Villalobos, M., Bargar, J., Sposito, G., 2005a. Mechanisms of Pb(II) sorption on a biogenic manganese oxide. *Environ. Sci. Technol.* 39, 569–576.

Villalobos, M., Bargar, J., Sposito, G., 2005b. Trace metal retention on biogenic manganese oxide nanoparticles. *Elements* 1, 223–226.

Webb, S.M., Tebo, B.M., Bargar, J.R., 2005a. Structural characterization of biogenic Mn oxides produced in seawater by the marine *Bacillus* sp. strain SG-1. *Am. Mineral.* 90, 1342-1357.

Webb, S.M., Dick, G.J., Bargar, J.R., Tebo, B.M., 2005b. Evidence for the presence of Mn(III) intermediates in the bacterial oxidation of Mn(II). *PNAS* 102, 5558-5563.

Weiner, S., Dove, P.M., 2003. An overview of biomineralization and the problem of the vital effect. In: Dove, P.M., Weiner, S., De Yoreo, J.J. (Eds.) Biomineralization. Mineralogical Society of America, Washington, DC, v. 54, pp. 1-31.

White, W.B., Vito, C., Scheetz, B.E., 2009. The mineralogy and trace chemistry of black manganese oxide deposits from Caves. *J. Cave Karst Stud.* 71, 136-143.

Wilderer, P.A., Characklis, W.G., 1989. Structure and function of biofilms. In: Characklis, W.G., Wildener, P.A. (Eds), Structure and Function of Biofilms, Wiley, Chichester, pp. 5-17.

Yan, D., Yan, P., Cheng, S., Chen, J., Zhuo, R., Feng, J., Zhang, G., 2009. Fabrication, in-depth characterization, and formation mechanism of crystalline porous birnessite MnO_2 film with amorphous bottom layers by hydrothermal method. *Cryst. Growth Des.* 9, 218-222.

Zhang, Z., Lian, B., Hou, W., Chen, M., Li, X., Li, Y., 2011. *Bacillus mucilaginosus* can capture atmospheric CO_2 by carbonic anhydrase. *Afr. J. Microbiol. Res.* 5, 106-112.

In: Advances in Geochemistry Research
Editor: Jorge Sanjurjo-Sánchez ISBN: 978-1-62618-245-5

Chapter 6

ADVANCES OF APPLIED GEOCHEMISTRY IN CHINA

Xie Xuejing, Cheng Zhizhong, Yao Wensheng, Cheng Hangxin and Sun Zhongjun
Institute of Geophysical and Geochemical Exploration (IGGE), CAGS, P.R. of China

ABSTRACT

The advances of Applied Geochemistry in China are summarized and briefly described in this chapter.

The Regional Geochemistry National Reconnaissance (RGNR) program initiated in 1979 has now covered about 7.7 million (80%) of China's territory by within-cell stream sediment sampling. The results will soon be published in an Atlas for the distribution of 39 elements in China both in English & Chinese. The tremendous amount of 39 element data has led us to establish the geochemical block concept for predicting world-class mineral deposits, and to the discovery of about 1900 economic mineral deposits.

A program of Geochemical mapping of 76 element (96GEM) using the composited samples of South & SW China derived from the RGNR sample bank was finished and will soon be published as an atlas both in English & Chinese.

The multi-purpose regional geochemical soil survey program initiated in 1999 had covered 1.65 million km^2 in Eastern China. After a time of postponement, it will now be renewed to cover whole China. The

Geochemical Atlas of Hainan Island, Panyang Basin of Jiangxi and Pearl River delta were published for environmental, agricultural and ecological studies.

A program of studying nearly all element contents in the periodic table in all the ore tailing dumps in China was initiated.

In order to overcome the general shortcomings of weak selective extraction used in Deep penetrating geochemistry (low contrasts of anomalies, and low precision of analysis) new approaches have been tested including: (1) ligand based extraction (2) strong acid extraction of micro-particles separated from samples (3) strong acid selective extraction of the stable end products of variable oxides & sulfides after the mobile unstable form of element are moved to the surface through exotic covers.

Large orientations studies for surface geochemical and microbiological surveys of oil and gas were initiated in Heilongjiang province including in region where the famous Daqing oil field located and also in Sichuan Province and Tibet. Geochemical surveys for oil and gas began to be used in China in the 1980's, and nearly stopped at the end of 20th century. A new large program of geochemical exploration for oil and gas to cover whole China was worked out. It could be started after the success of the orientation studies.

Chinese geochemists had been active in the international research program IGCP 259 for standardization of the multi-element analysis system, and IGCP 360 in the study of sampling methodology for the global-scale geochemical mapping. In view of the slow progress of global scale geochemical mapping, new concepts based on the fractal self-similarity principle were proposed. And an international research center will be established in IGGE to promote the research works after it will be approved by UNESCO.

1. INTRODUCTION

For the development of the Applied Geochemistry, China's main research activities have been focused to obtain the information of the spatial distribution of nearly all elements in the periodic table in various scales, which are of fatal importance in the exploration of mineral deposits and on the study of environmental and agricultural problems.

2. Regional Geochemistry, National Reconnaissance Program

The Regional Geochemistry National Reconnaissance program (RGNR) initiated in 1979 (Xie, 1979a; Xie 1989b) has now been covered about 7.7 million Km^2 (about 80% of China's territory) (Figure 2.1). By stream sampling and analysis of 39 elements, it makes great contribution to the discoveries of new mineral deposit and the science of geochemical mapping at that time. This program first advanced the idea that the detection limits of all trace & sub-trace elements should be lowered down beneath the crustal abundance (Xie, 1979a; Xie 1979b). For this idea to be fulfilled, China insisted in all these 30 years to develop and improve multi-instrument multi-method analytical systems (Xie, Cheng and Liu, 2008). For obtaining the enormous data in cost effective way, China has divided the analytical procedures into large scale stream line operations inspired from Ford's innovative idea in the industry for production of automotive cars (Qian and Xie, 1953; Xie, 1995).

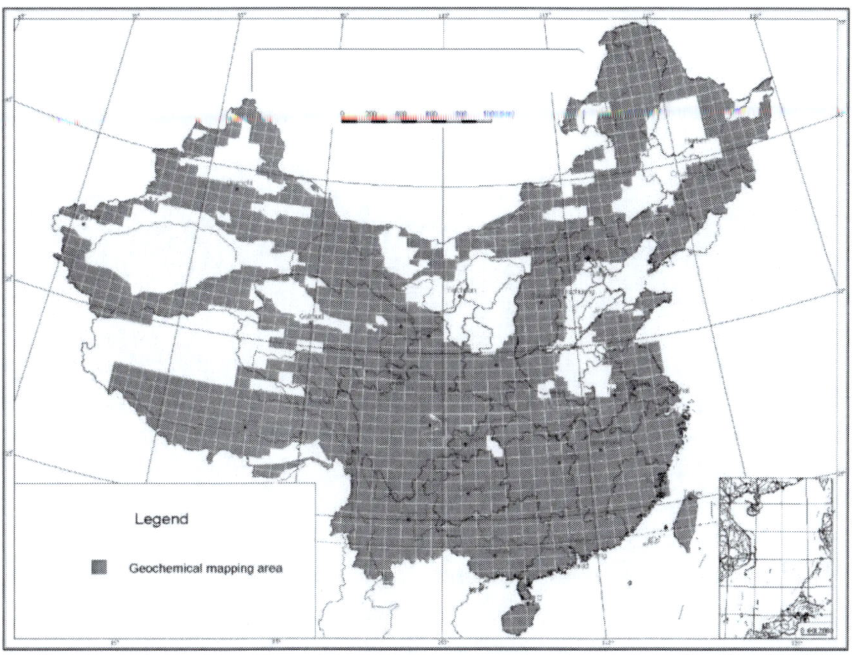

Figure 2.1. The coverage of RGNR program till 2012.

Due to the complex landscapes in China (Figure 2.2), Chinese geochemists have developed different sampling methods for different landscapes (Xie & Ren 1991). For China proper landscape (mountain and hilly) standard within-cell stream sampling with sampling density 1-2 samples/km^2 was used. Composite of samples from four 1km^2 cell were undertaken for analysis (Xie 1977).

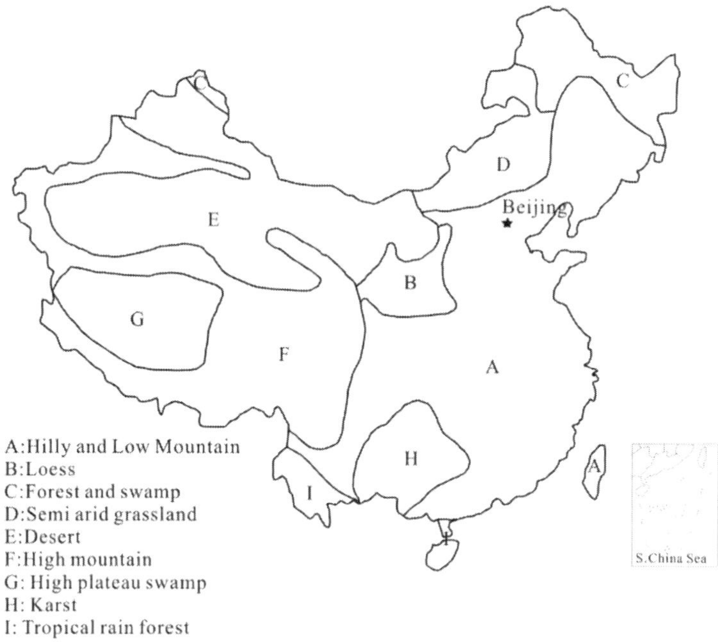

Figure 2.2. Geographic landscapes in China.

It was predicted at the initiation of the RGNR program that more than tens thousand regional anomalies will be delineated, more than thousands anomalies will be further explored and the results will be more than hundreds economic deposits discovered (Xie et.al, 1986). Now thirty years has elapsed, 1900 economic mineral deposits were discovered (mainly gold, silver, base metals, copper, tungsten, tin and molybdenum deposits) (Xie et al., 2011a) (Figure 2.3) with the values of potential ore reserve exceeding 1600 billion Yuan. This fact is far more than the prediction at the initiation stage of the program (Xie et. al, 2011a; personal communications with Mao and Wang, 2012).

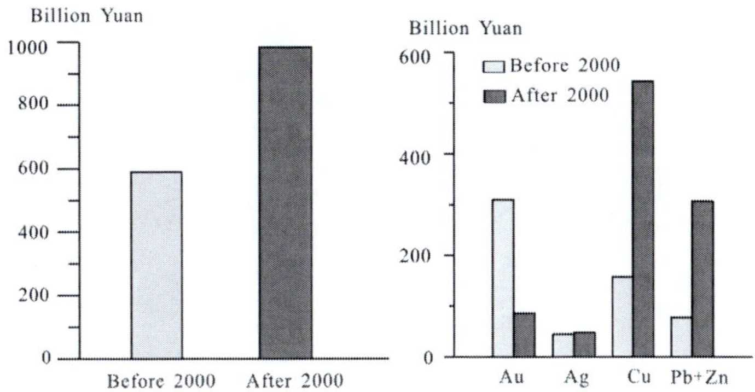

Xie, et al., 2011a.

Figure 2.3. Assessment of economic values of discoveries by the RGNR project according to their potential reserves.

The enormous amount of element data obtained provided the basis for the formation of many new ideas and concepts in applied geochemistry such as the hierarchy of geochemical patterns from local, regional, provincial mega-provincial, dominical, continental to global, the geochemical block concept for searching the large ore deposits and the corresponding sampling techniques from detailed to global scale (Xie & Yin 1993, Xie et al., 2004).

The multi-element analytical system developed for RGNR program has been continuously sophisticated and enlarged during these 30 years (table 2.1). Standard reference materials GSD1-8, GSR1-6 and GSS1-8 were prepared and certified (Xie et al., 1985; Xie et al., 1989a). The analytical data quality monitoring procedure was worked out and used though these 30 years(Xie, 1979b).

Table 2.1. Analytical scheme of RGNR for 39 elements

X-ray fluorescence spectrometry	Al, Ba, Ca, Co, Cr, Cu, Fe, K, La, Mg, Mn, Na, Nb, Ni, P, Pb, Si, Sr, Th, Ti, V, Y, Zn, Zr
Atomic fluorescence spectrometry	As, Sb, Bi, Hg
Atomic absorption spectrometry	Ag, Cd, Li
Atomic emission spectrometry	Ag, B, Be, Sn
Catalytic polarography	W, Mo
Ionic selection electrometry	F
Colorimetry or laser catalytic fluorescence	U
Graphite furnace atomic absorption spectrometry	Au

Xie, 1995; Darnley et al., 1995.

The Atlas of the distribution 39 elements will soon be published (Figure 2.4).

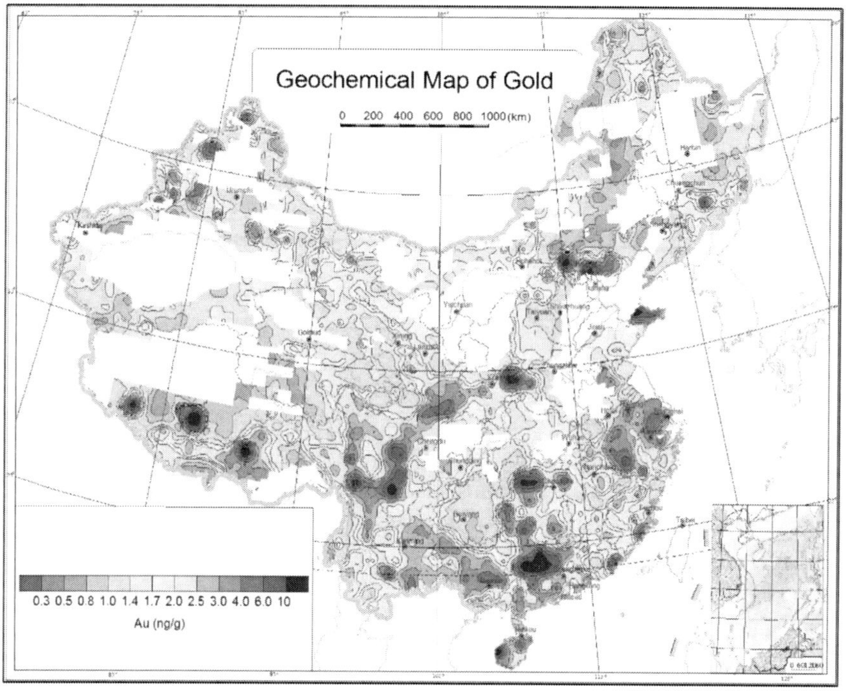

From Xie et al., in print.

Figure 2.4. Geochemical map of gold in China.

3. THE 76 GEOCHEMICAL ELEMENT MAPPING PROGRAMS (76 GEM)

The 76GEM program was proposed in 1999. The aim of this program is to analyze and study of the distribution of most elements in the periodic table by using the composite samples taken from the RGNR data bank. In addition to the 39 elements analyzed in RGNR program, 37 more elements were analyzed in the composite samples (Table 3.1). The development of analytical methods of these additional 37 elements has been continuously all along before and during this program in the study and preparation of China's standard reference samples GSD, GSR and GSS series.

Table 3.1. Analytical elements for 76 elements geochemical mapping program

39 elements Determined by RGNR	Al_2O_3, CaO, Fe_2O_3, K_2O, MgO, Na_2O, SiO_2, Ag, As, Au, B, Ba, Be, Bi, Cd, Co, Cr, Cu, F, Ga, Hg, La, Li, Mn, Mo, Nb, Ni, P, Pb, Sb, Sn, Sr, Th, Ti, U, V, W, Y, Zn, Zr
Newly determined elements	Rare and dispersed elements: Cs, Ga, Ge, In, Rb, Re, Se, Te, Tl Volatile element: Hf, Ta, Sc Halogen elements: Br, Cl, I Platinum group elements: Pd, Pt, Ir, Os, Rh, Ru Rare earth elements: Ce, Pr, Nd, Sm, Eu, Gd, Tb, Dy, Ho, Er, Tm, Yb, Lu Non-metallic elements:N, S, C

The first subprogram was initiated in 2000 to cover SW China including Sichuan, Yunnan, Guizhou and Guangxi (Xie et al., 2008a). One hundred samples within each 1:50,000 map sheet from RGNR sample bank were mixed together to prepare a composite sample. About 3000 composite samples (Figure 3.1) were submitted for analysis of 76 elements by multi-instrumental multi-analytical methods (Table 3.2). Instead of the "Standard reference samples" concept for monitoring analytical data quality, a new concept "reference maps" was advanced in the research work of this subprogram. One hundred and fifty reference samples were prepared by mixture in various proportions of GSD and GSS samples. These samples were arranged in virtualized spatial distribution. Their element values were calculated by the certified values of various proportions of original standard reference samples GSD & GSS series (Xie 2008b). Then "reference maps" were prepared, These maps was compared with maps prepared by element values of these reference samples analyzed by laboratories which will participated in the 76 GEM program. The similarity of maps with the reference maps was obtained by visual observation or by calculation of the statistical parameters. In the early stage this method was used for selection of competent laboratory to participate in this program. Later it has become the routine procedure for evaluation of the map quality in China. Figure 3.2 shows the comparison of the reference map of Cu with the same Cu map of laboratory A. Visual comparison shows the good similarity and the statistic parameters also show the good similarity. Figure 3.3shows the comparison of the reference map of Cl with the same Cl map of laboratory A. The visual comparison reveals that there are series systematic errors in map by laboratory A, and it also can be seen from the comparison of statistical parameters. Figure 3.4 shows the comparison of the reference map of Th with the same map based on the analytical data by laboratory A. Visual comparison reveals the map quality of laboratory A is not good, though the statistical correlation of the two sets of data is high up to 0.991 and the analytical precision also is very good.

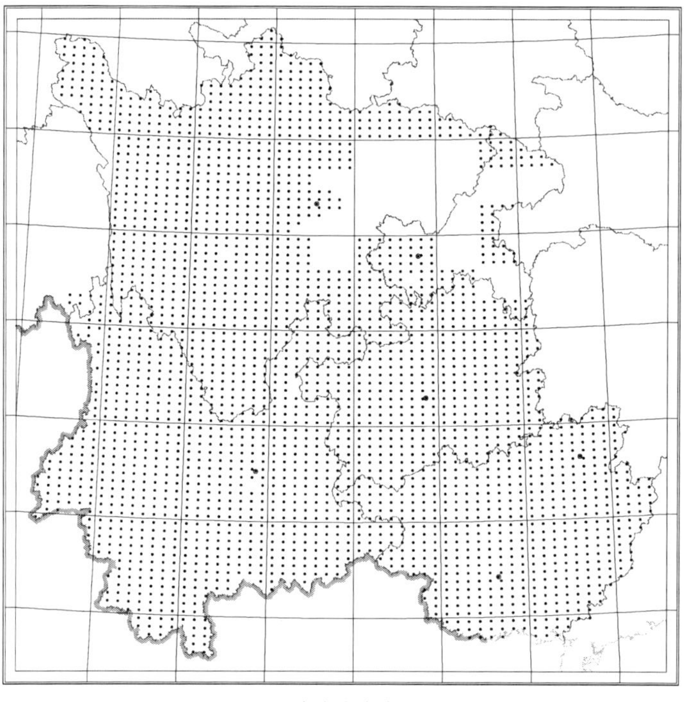

Figure 3.1. Distribution of composite samples in SW China.

The Atlas of 76 GEM of SW China were published in Chinese in 2008 (Xie et al., 2008b) later another sub-program 76 GEM in South China was initiated in 2007 which covered Guangdong, Fujian, Anhui, Jiangxi, Hubei, Hunan, Zhejiang, Jiangsu, Shandong provinces. An Atlas of 76 element distribution of both SW and South China was prepared both in Chinese and English (Figure 3.5, 3.6), will soon be published.

Table 3.2. The Analytical Scheme for 76 elements

Determined Method	Elements
XRF	SiO_2, CaO, TFe_2O_3, Ba, Br, Cl, Mn, Nb, P, Pb, Rb, S, Sr, V, Zn, Zr
DF-ICP-AES	Al_2O_3, K_2O , MgO, Na_2O, Be, Cr, Cu, Li
DF-ICP-MS	Bi, Cd, Co, Cs, Cu, Ga, Hf, In, Mo, Ni, Sb, Sc, Ta, Te, Th, Tl, U, W,
FU-I-ICP-MS	La, Ce, Pr, Nd, Sm, Eu, Gd, Tb, Dy, Ho, Er, Tm, Yb, Lu, Y
FU-I-ICP-MS	Re
DA-GFAAS	Au
HG-AFS	As, Se, Ge
CV-AFS	Hg
ISE	F
AES	Ag, B, Sn
VOL	C, N
CF-COL	I
DA-ICP-MS	Pt, Pd, Au
FU-ICP-MS	Ru, Os
FA-ICP-MS	Ir, Rh

Decomposition and Preconcentration Method: DF: Digestion with aqua regia and HF; DA: acid digestion; FA: Ni-S-fire assay; FU: alkali fusion.

Xie et al., 2008b.

Determination Method: XRF: X-ray fluorescence spectrometry; ICP/MS: inductively coupled plasma-mass spectrometry; ICPAES :Inductively couple plasma atomic emission spectrometry; GFAAS: graphite furnace atomic absorption spectrometry; HG-AFS: Hydride generation-non-dispersion Atomic fluorescence spectrometry; CV-AFS: Cold mercury vapour-non-dispersion Atomic fluorescence spectrometry; AES: atomic emission spectrometry; COL: Colorimetry, VOL: Volumetry; ISE: Ion-selective electrode.

Reference map of Cu Same map of Cu by Lab A

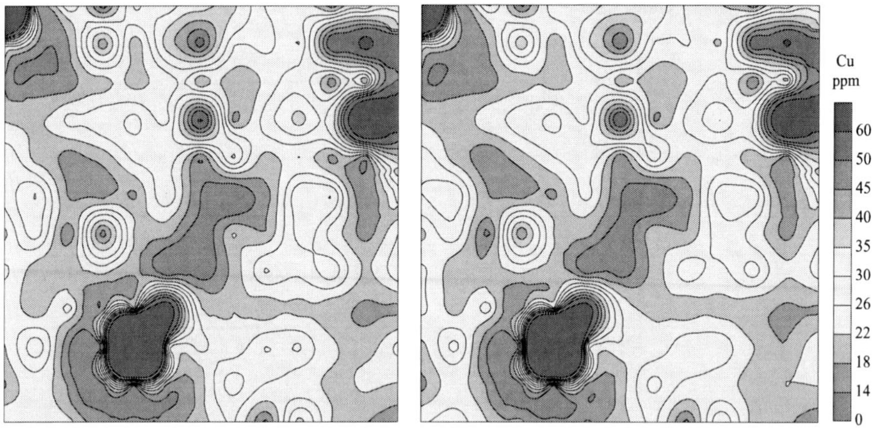

Reportability(%)	100
Precision	
Average RSD(%)	6.92
RSD acceptable/Total	150/150
Accuracy	
Average ΔlogC	0.02
ΔlogC acceptable/Total	150/150
Mean value (ppm) of whole data set	
1. Measured	27
2. reference	28
Medium value (ppm) of whole data set	
1. Measured	22
2. reference	22
Variation (%) of Whole data set	
1. Measured	69.6
2. reference	63.7
Correlation	0.999

Figure 3.2. Visualized Cu maps (2) produced by reference data and XRF data by Lab A.

Reference map of Cl Same map of Cl by Lab A

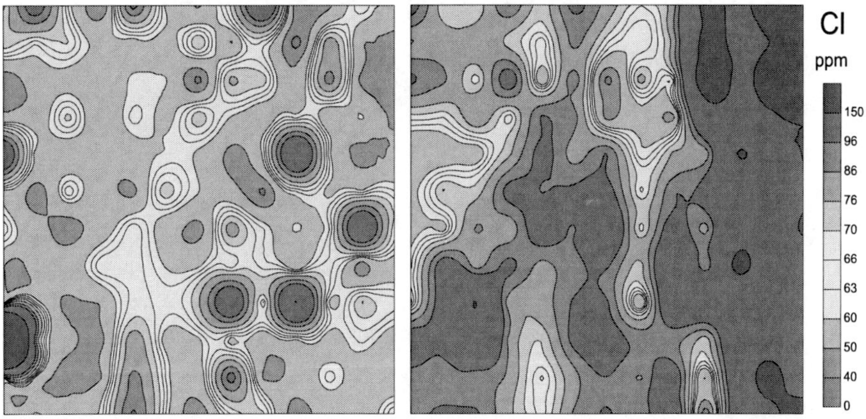

Reference map of Cl	Same map of Cl by Lab A
Reportability(%)	100
Precision	
Average RSD(%)	22.3
RSD acceptable/Total	80/150
Accuracy	
Average ΔlogC	0.11
ΔlogC acceptable/Total	38/150
Mean value (ppm) of whole data set	
1. Measured	97
2. Reference	66
Medium value (ppm) of whole data set	
1. Measured	94
2. reference	59
Variation (%) of Whole data set	
1. Measured	33.9
2. reference	62.0
Correlation	-0.034

Figure 3.3. Visualized Cl maps (2) produced by reference data and analysis data by Lab A.

Reference map of Th Same map of Th by Lab A

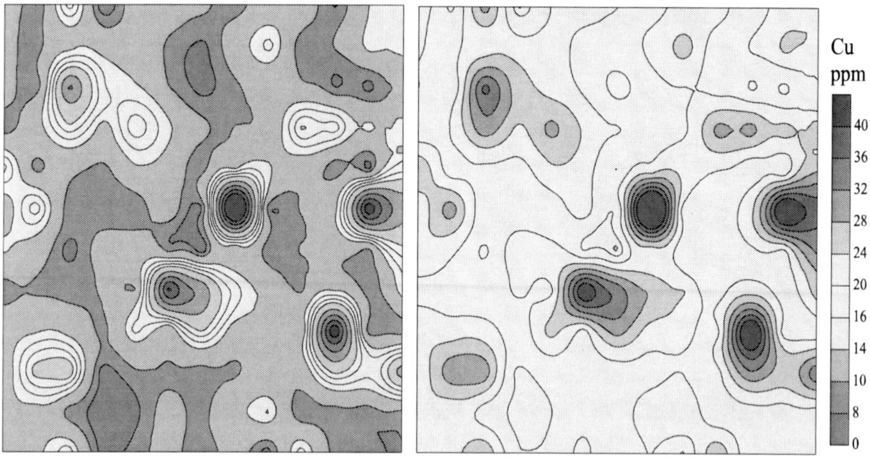

Reportability (%)	100
Precision	
Average RSD(%)	6.67
RSD acceptable/Total	150/150
Accuracy	
Average ΔlogC	0.25
ΔlogC acceptable/Total	3/150
Mean value (ppm) of whole data set	
1. Measured	21.7
2. reference	12.7
Medium value (ppm) of whole data set	
1. Measured	20.2
2. reference	11.1
Variation (%) of Whole data set	
1. Measured	33.6
2. reference	53.6
Correlation	0.991

Figure 3.4. Visualized Th maps (2) produced by reference data and analysis data by Lab A.

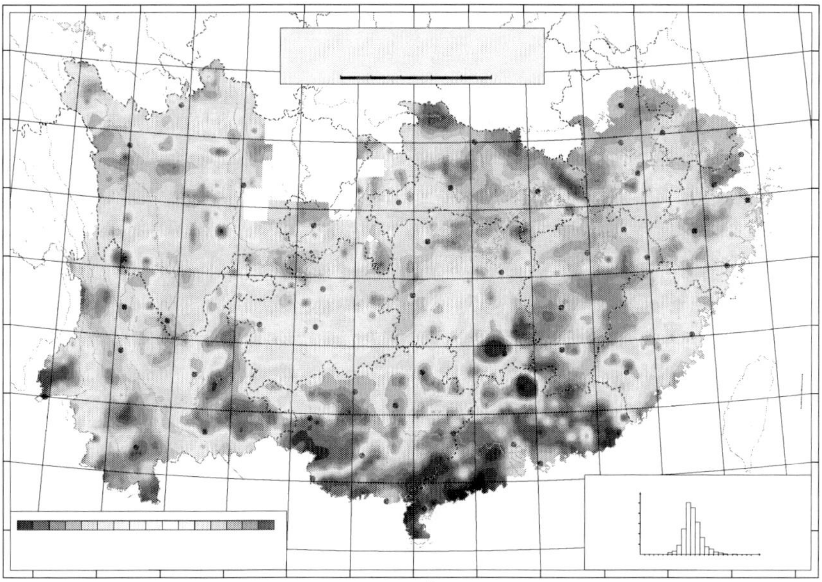

Cheng & Xie, in print.

Figure 3.5. Geochemical map of Tellurium (Te) in SW and S China.

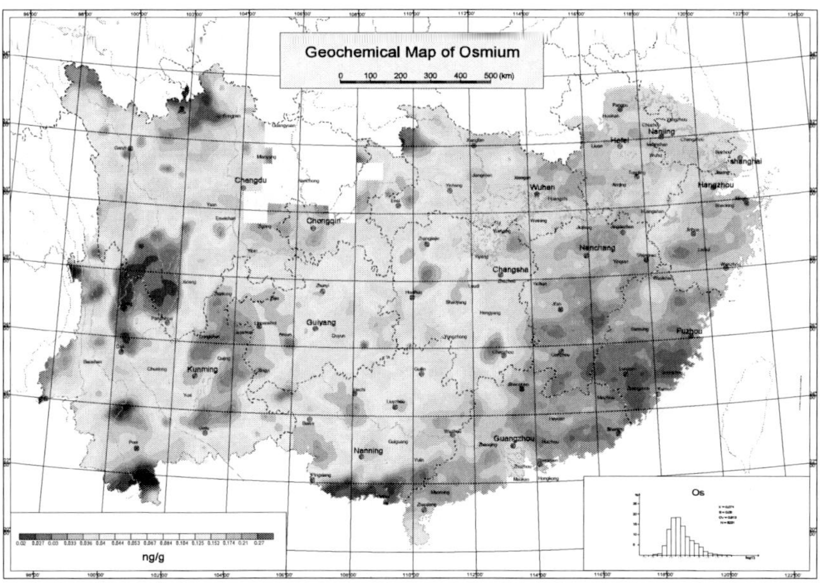

Cheng & Xie, in print.

Figure 3.6. Geochemical map of Osmium (Os) in SW and S China.

4. The Multi-purpose Geochemical Surveys

The content of this section is summarized from manuscript of Cheng Hangxin: "Multi-purpose Geochemical Surveys - present and future development (in Chinese, unpublished) The Multi-purpose Geochemical Surveys had been initiated in 1999. One surface soil sample was taken in each one km^2 cell and one soil sample at depth of 1m was taken in each 4 km^2 cell Composite surface sample from 4 one km^2 cell and composite depth samples from 4 four km^2 cell were prepared and submitted for analysis of 54 elements. The program had been carried out for 9 years and covered 1.65 millions km^2 of China's Eastern part. The aims of this program were to study the ecological, environmental and agricultural problems. Based in the ecological units : farmlands, drainage systems, urban regions, lakes, wetlands, shallow seas and grass plains, the geochemical element patterns were delineated. Further sampling of vegetation, soil air, atmosphere, suspended particles in air, agricultural products, river sediments lake sediments, suspension in streams, stream waters etc were taken and studied in problem areas, In order to understand the pollution sources, pollution paths, and their ecological effects. Table 4.1 shows the soil quality expressed as the abundance of essential elements obtained by the above studies.

Table 4.1. The plentifulness and shortness of some essential element in China

Essential elements	Shortness (%)	plentfulness (%)
P	78.69	21.31
N	18.00	82.00
K	0.71	99.29
Organic C	13.89	

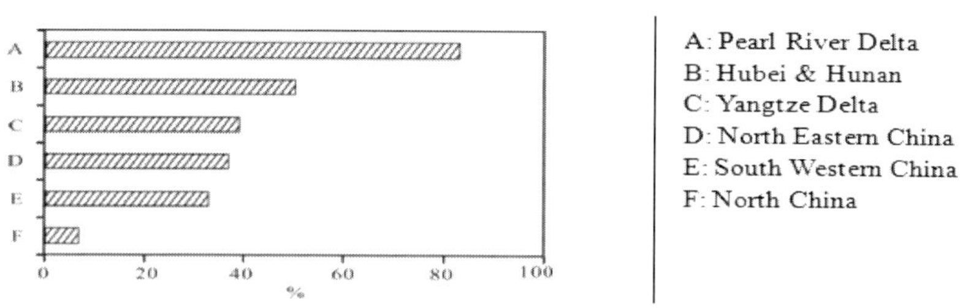

Figure 4.1. The quality of soil expreseed by the Percentage of acidification (pH 5.5-3.5) in different parts of China.

As for the pollution of farmland soil in China Cd pollution (0.3-1.5ppm) is series along Yangtze River with areal extent to 13,000 km^2. High values of As and F are along the Yellow river, especially some poor & remote villages are suffered with various kinds of diseases caused by anomalous As and F.

Dry and wet atmospheric precipitation will be the main path of pollution in farmlands, The pollution by irrigation water and fertilizers are 10.48% and 4.32% respectfully.

Mercury, lead and arsenic pollution in urban regions are remarkable, especially mercury (Figure 4.2) Radioactive pollution are found in some Urban regions.

Bottom sediment in lakes in China are series: Sn, Cu, Cd, Pb pollution in Boyang Lake, Cd, Hg and S in Tai Lake, Pb, Cd in Dongting Lake, As, Sb, Cd in Hongze Lake, and Cd in Chao Lake.

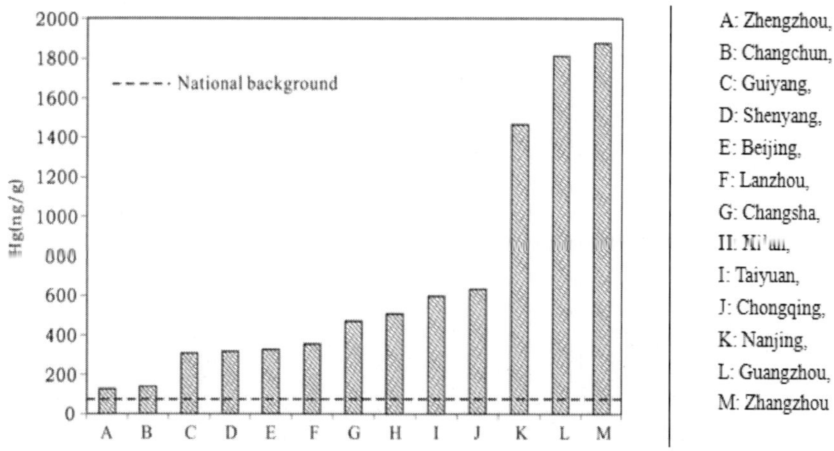

Figure 4.2. Average concentration of Hg in urban soil in China.

Mercury in soils of urban regions is mainly in the form of HgS. Experiments showed that HgS in Soil could be vaporized into air. Figure 4.3 shows the positive correlation of the content of HgS in soil with Hg in air Figure 4.4 shows the positive correlation of certain amount of HgS with Hg in air at various temperature.

Figure 4.5a shows the variation of HgS in Beijing before 1987. Figure 4.5b shows that after 1987, the government has made great effort to replaces coal burning for gas and HgS in soil began to drop down in 2008. It could be postulated that Hg in air were also drop down.

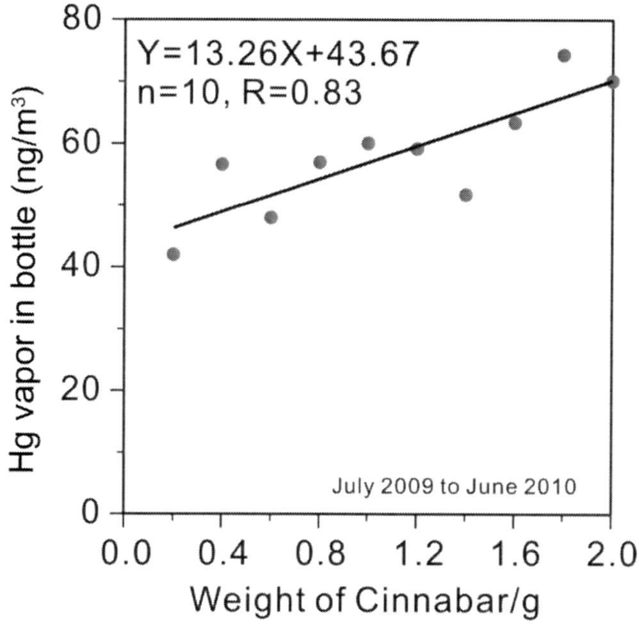

• Average value of Hg in air with 12th determinations from July 2009 to June 2010.

Figure 4.3. Positive correlation of content of HgS in soil with Hg in air.

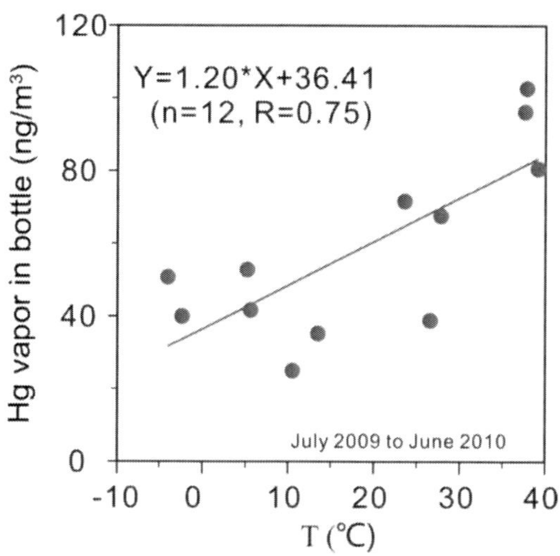

• Average of Hg in air with 12th determinations at a cetain temperature from July 2009 to June 2010.

Figure 4.4. Positive correlation of content of HgS with Hg in air under different temperature.

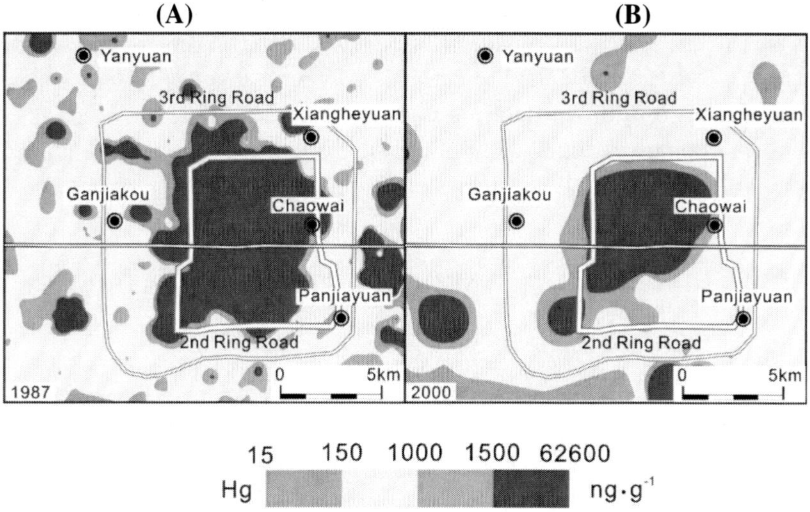

Cheng et al., 2008.

Figure 4.5. Mercury (Hg) distribution in soils of Beijing city in 1987 and in 2000.

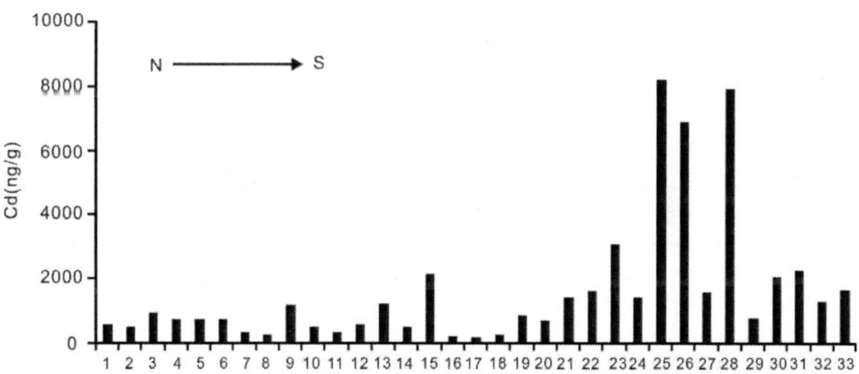

1-Liaohe river, 2-Shuangtaizihe river, 3-Dalinghe river, 4-Luanhe river, 5- Haihe river, 6-Mayinghe river, 7-Tuhehe river, 8-Huanghe river, 9-Xiaoqinghe river, 10-Xinyihe river, 11-Subei river, 12-Sheyanghe river, 13-Yangze river, 14-Huangpujiang river, 15-Qiantangjiang river, 16-Caoejiang river, 17-Jiaojiang river, 18-Oujiang river, 19-Jiaoxi river, 20-Mengjiang river, 21-Jinjiang river, 22-Jiulongjiang river, 23-Hanjiang river, 24-Dongjiang river, 25-Beijiang river, 26-Xijiang river, 27-Tanjiang river, 28-Muyangjiang river, 29-Jianjiang river, 30-Nanliujiang river, 31-Qinjiang river, 32-Changhuajiang river, 33-Wanquanhe river.

Figure 4.6. shows the sediment Flux of Cu from tributary draining into Yangtze River.

This program was temporally stopped in 2008. In 2012 a more extensive program covering whole China will be renewed. Due to data of carbon were obtained in the past in RGNR Project, EGMON project and MPRGS project in Eastern China so in the present program, loss or gain of carbon in farm land and other ecological units could be systematically studied and improved agricultural operations will be suggested.

5. Study of the Composition of Ore Tailings in China

The Mining Activities especially in recent 30 year have left ore tailings totally more than 20 billion tons. Due to the past approach of mining, only the main ore element was extracted and many valuable trace elements were left in the tailings.

Table 5.1. The concentration of element in some Tailing dumps

Tailing	Element	Concentration
Dexing copper deposit	Cu	0.10%
Shizhuyuan polymetallic deposit	Sn	0.10%
	W	0.12%
	Ge	9.4μg/g
	In	7.3μg/g
	Te	3.0μg/g
Hongqiling nickel deposit	Ni	0.31%
	Se	5.0μg/g
Daye copper deposit	Cu	0.20%
	Se	5.6μg/g
Gejiu tin deposit	Fe	19.6%
	Sn	0.23%
	Pb	0.72%
	Zn	0.36%
	Cu	0.20%
	In	13.3μg/g
Dajishan tungsten deposit	WO_3	0.13%
Lingshuijiang antimony deposit	Sb	0.43%
Huangshapin lead and zinc deposit	Fe	20.3%
	S	16.1%
	Zn	0.55%
	Tl	34.0μg/g
	Te	2.6μg/g

Cheng, unpublished.

Many cities or regions have long history of mining only the main ore element (Copper, base metals, gold, tungsten, tin or molybdenum) and have kept and economic prosperity for a long period of time, but now faced economic depression due to the exhaustion of main ore reserve. An amount of main ore element or other more valuable rare or rare earth elements still left in the large dumps of ore tailing and economic prosperity will be renewed, when these could be utilized with more innovative techniques.

For this the first important step will be a national survey to know the composition of all the tailing dumps in China by geochemical method. Orientation study was already carried out in 2010 (table 5.1).

6. Deep-Penetrating Geochemistry

Tradition geochemical exploration method is less efficient in discovery of ore bodies concealed beneath various kinds of transported overburdens (alluvial, wind-born, or volcanic overburdens). So increasing attention is being focused on the development of deep-penetrating geochemistry. Numerous publications have documented by using weak selective extraction to detect only the mobile form of elements that could penetrate through the exotic layer and filter out the noise of background variation that could mask the weak anomalous value of mobile form of ore element. Such methods have been extensively used by industry abroad with some successes for finding concealed deposits and many dry holes (Cameron 1998). The shortcomings of selective weak extraction are (1) very low contrast of anomalies with background values (2) low precision, caused by variation of sample type, working environment and personal equation. Study now in China has provided several new approaches to overcome these shortcomings: (1) strong selective extraction to detect the stable end products after the mobile unstable forms came up in contact with air. This experiment was successfully carried out in Spence porphyry deposit in Chile (Figure 5.1) (Xie et al. 2011b). (2) strong acid extraction of selective grain size samples (samples taken at 0-40cm depth and sieved to <160 mesh) (Wang and Ye 2011). These fine particle samples are the main absorbants of the mobile elements carried up through exotic overburdens to the surface. (3) Strong legand based soil extraction which could limit matrix element interference, reduce background noise and overcome PH effect. This method was first advanced by Mann (Mann2010) and improved by Yao (Yao, personal communication 2011). A new orientation project was initiated to test the above new approaches of deep penetrating methods in cooperation with overburden drilling. The studies could both testify the effectiveness of any new approach of deep penetrating method and improve the efficiency of Chinese-made overburden drills. After that a big science program will be proposed to cover all the difficult regions in China: regions covered with alluvial, wind-born and volcanic overburdens and also loess covered regions.

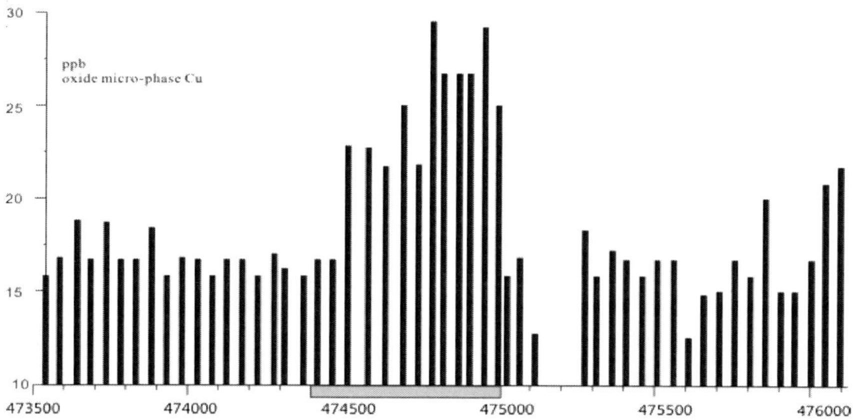

Figure 6.1. Strong selective extraction of micro-phase oxide Cu from exotic cover along one traverse across concealed Spence porphyry copper deposit, Chile (from Xie et al., 2011b).

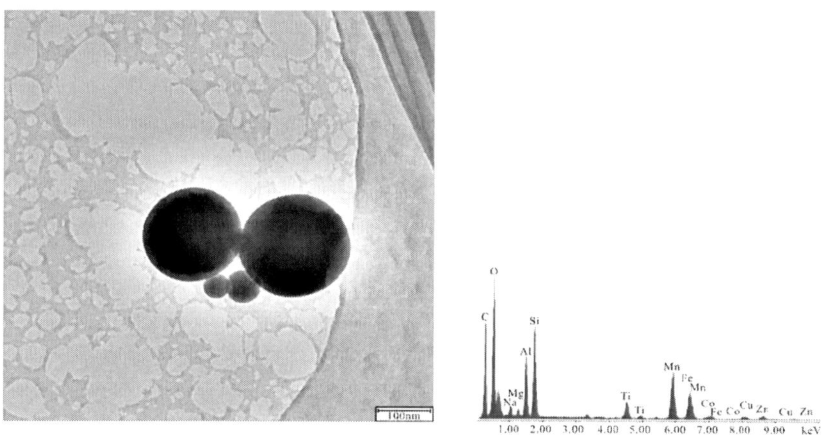

Figure 6.2. Nanoscale metal particles of Si-Al-O-Mg-Fe-Mn-Ti-Cu-Co observed by TEM over a Cu-Ni deposit concealed at a depth of 400 meters in the verge of Nanyang basin, Henan.

Beside the technologies of deep-penetrating geochemistry, studies of the mechanism and form of mobile elements were also carried out. Xie discussed the relayed effect of various driving agencies with earth gas as the main driving force relayed or reinforced by seismic pumping, barometric pumping, CO_2 generator, evaporation, capillarity action and plant root action in various environments (Xie and Wang 2003). Wang & Ye (2011) discovered that the so called mobile form of ore elements will be in nano-scales in exotic overburdens over mineral deposited through TEM observations (Figure 6.2).

In order to understand the existence and migration of nano-scale particles though exotic covers, a migration pillar 250*250cm was prepared in the laboratory, on the bottom of the pillar 50cm ore particle were filled covered by 200cm soil. After 16 month nano-particle of Cu-Fe (n to n*10 nm) were observed in the surface layer of soil (Figure 6.3).

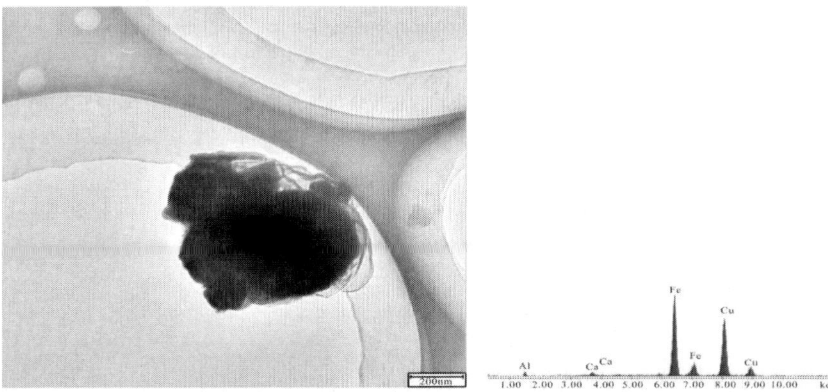

Figure 6.3. Nanoscale metal particles of Cu-Fe observed by TEM.

7. Surface Geochemistry of Oil and Gas Exploration

The development of surface Geochemistry in oil and gas exploration in China just as the same in other countries has been full of troubles and difficulties. Unlike the geochemistry for mineral exploration, whose concept and application of dispersion patterns of ore elements were quickly accepted, the concept and application of vertical migration of hydrocarbons through well-sealed oil or gas pool seems absurd and difficultly to be accepted by most petroleum geologists. Besides, seismic technology now pervades the industry and has advanced so that it is quite capable of finding even subtle traps. Another factor that hindered the development of surface geochemistry for oil and gas exploration in China was the lack of strategic big science program like the RGNR program in mineral exploration (Xie 2009). In the past there has been only many small subprojects subordinated to large programs of petroleum geology or seismic survey in a beggarwise way. Their contributions were often neglected. A big science program for surface geochemistry in Oil and Gas exploration was suggested in 2009 (Xie & Sun 2009) The program includes the covering of whole China's stratagraphic basins (Figure 7.1) by a multi-method approach (soil gas, △C, radiometric, fluorescence, mercury, halogens, major and minor elements, microbiological methods) at a sampling density of 1 sample/25km^2 for basins<10^4 km^2 and 1 sample/100km^2 for basins>10^{4}km^2. Orientation surveys have already initiated in Heilongjiang and Tibet with a sampling density of 1 sample/ 25km^2 and semi-detailed orientations in the surroundings of Daqing Oil field (Figure 7.2).

Figure 7.3 and Figure 7.4 show the good result of ΔC and Sr distribution in Heilongjiang region which encourages us to believe that the strategic role of surface geochemistry for predicting favorable producing basins at more cost-effective way than seismic method especially in topographic difficult regions.

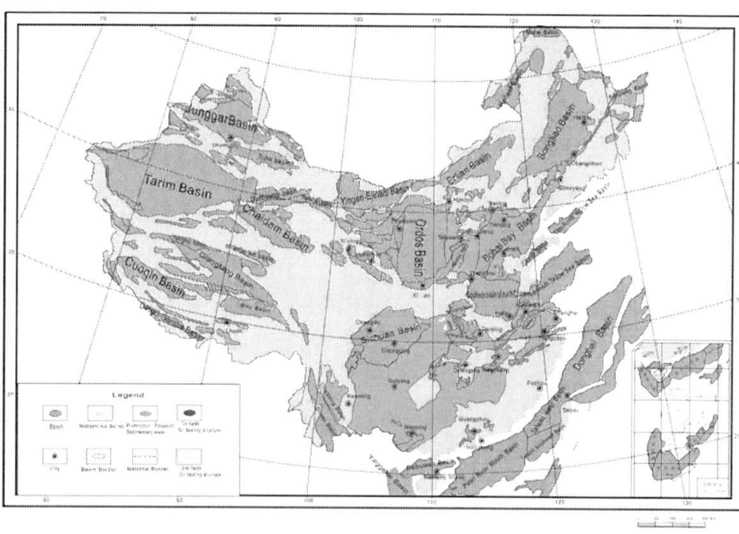

Figure 7.1. Map of China's strata graphic basins.

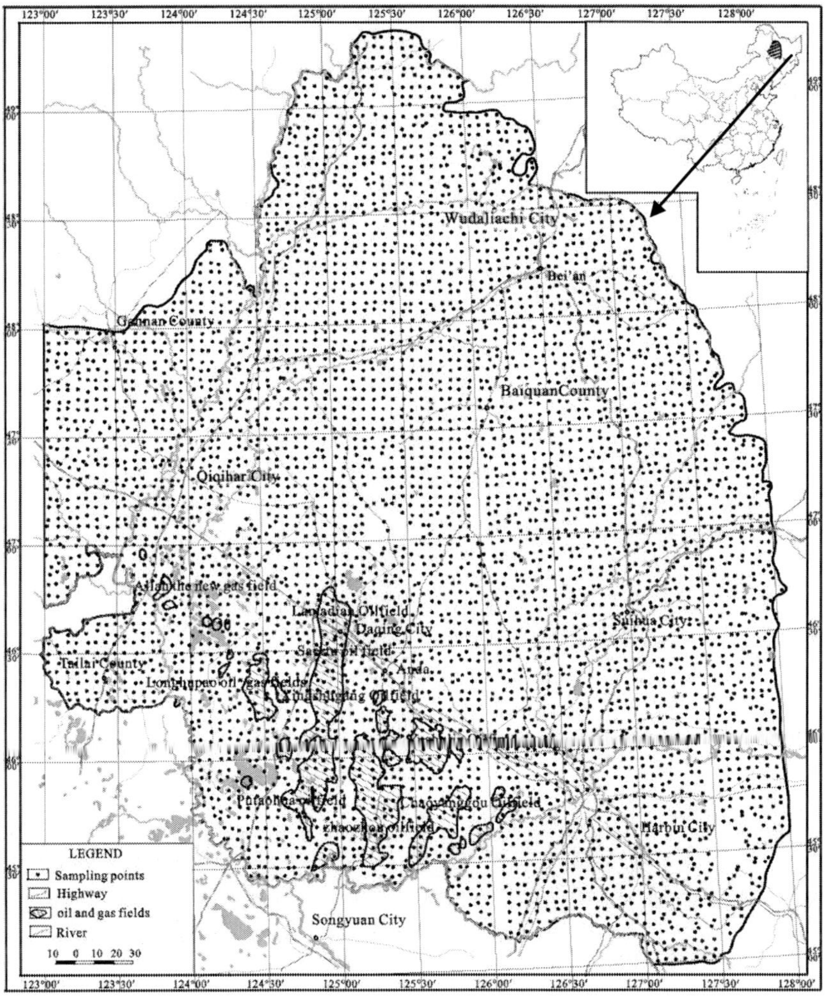

Figure 7.2. The sampling site of geochemical mapping in Songliao basin.

In addition, a new study of surface geochemistry was carried out for testing the effect to search buried hydrocarbon hydrate in west China (Figure 7.5).

Besides this strategic role, surface geochemistry can provide the clue whether the structures discovered by seismic survey is filled with oil and gas or not. Such kind of detailed surreys are now conducted by Wuxi petroleum research institute and private companies like AEAE Geo microbial Technologic. Figure 7.6 show an example of AEAE's success in locating a productive drill hole.

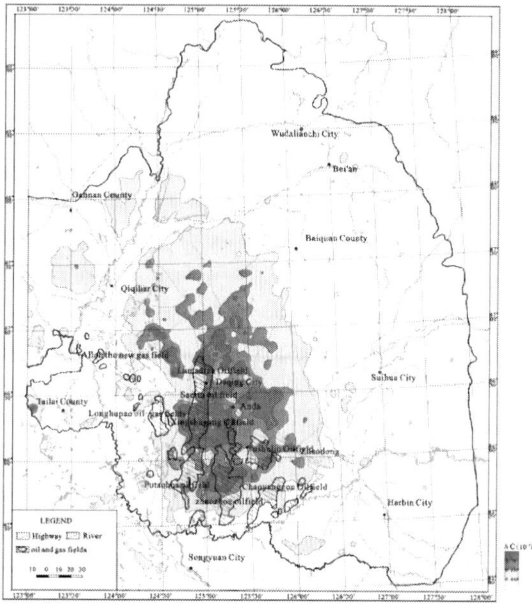

Figure 7.3. ΔC anomalies map in Songliao basin.

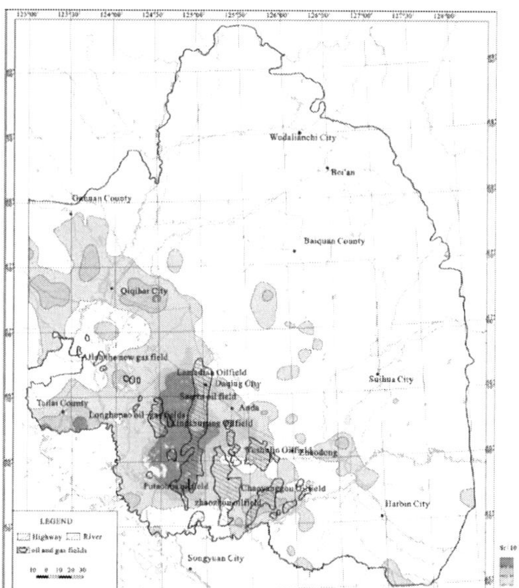

Figure 7.4. The map of Sr anomalies in Songliao basin.

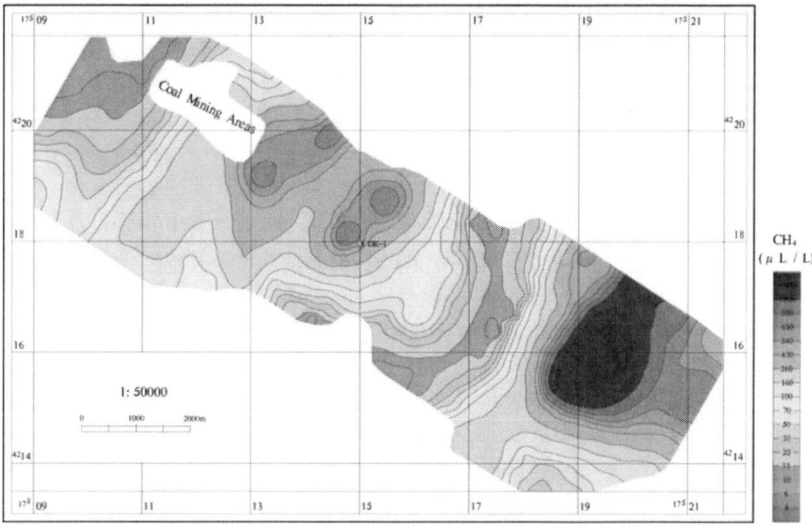

Figure 7.5. Hydrocarbon hydrate geochemical map (DK-1 hydrocarbon hydrate drill.

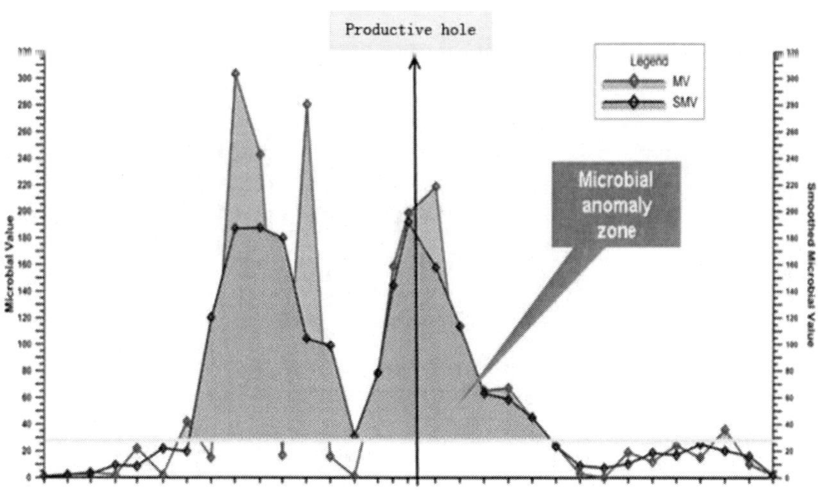

By courtesy of Mei Bowen.

Figure 7.6. Microbiological anomaly over Eastern Junggar Basin.

8. Global Scale Geochemistry

It's the dream of many geochemists in the world to get a global picture of nearly all the elements in the periodic table. If this dream realizes, it will be a big event after the discovery of periodic table by Mendeleev. China pioneered to use about 500 floodplain sediment composite samples to cover whole China (Xie and Cheng 1997; Chang et al., 1997). Later Bolviken (Norway) Bjorklund (Finland) and Xie (China) made a proposal to have 5000 sampling cells (each 160 X160km) to cover the whole world land surface (Figure 8-1). This proposal became the content of IGCP 360 (1992-1997) and IUGS/IAGC'S Task Group on Global Geochemical Baseline (1998-). More than 10 years have now elapsed, in addition to China, only Europe, India and Australia have completed their Baseline programs. It is difficult to carry out the same kind of works in S America, Africa and most part of Asia because the lack of funding and the inaccessibility of many cells (160X160km^2) in these countries due to political, cultural and social reasons. In order to get a global picture of most element distributions, new mapping concept should be worked out and a permanent agency should be responsible to foster the global mapping work with new mapping concept.

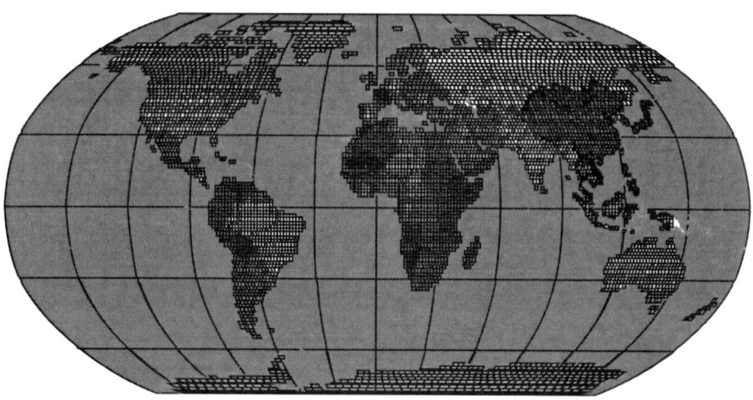

Based on the cells coordinate data from www.globalgeochemicalbaselines.eu

Figure 8.1. The 5000 GRN cells covered the Earth land area.

Bolviken and Xie advanced the idea that distribution of element in drainage system will follow the concept of fractal self-similarity (Bolviken 1992, Xie and Yao 2010). It means under the similar natural processes: rocks or soils eroded into stream will be well mixed and transported by water. So, samples taken at the month of drainage basin with various dimensions, these element contents will approximate the average element values of the respective upper drainage basins from several km^2 to hundreds, thousands or even millions of km^2 through the long span of geological time. The encouraged results by Xie and Tong made at the mouth of main tributary of Yangtze rivers （Table 8.1） and at the mouth of whole Yangtze river (Table 8.2) made us to advance a new program of global mapping (Xie and Yao 2011). This program will be initiated after the approval by the Chinese government and the UNESCO.

Table 8.1. Comparison of results of twice sampling of floodplain sediment

		W	Sn	Sb	Pb	Zn	Cu	Ag
Xiang River	SS	6.31	8.45	3.93	39.7	87	30.0	103
	1	6.5	9.67	4.22	55.1	133	31.3	141
	2	5.5	10.4	3.55	62.0	125	34.6	147
Reserve×10000t		147	92.2	10	425	618	60	0.9
Zi River	SS	3.55	6.18	12.9	32.5	91	28.8	76
	1	2.94	4.92	10.15	31.7	97	26.0	85
	2	2.57	4.1	13.47	31.0	81	27.9	94
Reserve×10000t		19	37	220	23	76	--	0.1
Yuan River	SS	2.39	3.56	3.27	25.5	91	28.3	104
	1	2.84	3.11	2.49	30.5	104	25.6	113
	2	2.24	3.28	4.32	31.9	92	27.6	107
Reserve×10000t		9.4	--	--	24	265	--	--

Note: SS average of whole basin (in ppm)
1 average at the mouth of river (in ppm) (1st sampling)
2 average at the mouth of river (in ppm) (2nd sampling)

Table 8.2. Comparison of the average element values of Yangtze delta floodplain sediments with average element values of the whole Yangtze basin region. (Ag in ng/g, others in ug/g)

	No.Data	**Ag**	**Au**	**Hg**	**Co**	**Cu**	**Li**	**Mn**	**Mo**
Yantyze	694817	102.0	2.4	107.0	15.7	33.4	41.2	810.0	1.3
Zhenjiang	17	106.5	1.6	80.5	15.6	31.9	42.2	789.6	0.7
Zhangjiagang	12	105.3	1.6	76.6	16.7	32.7	42.2	858.2	0.7
		Nb	Ni	Pb	Sb	Sn	U	W	Zn
Yangtze		19.0	34.6	33.1	1.7	4.4	3.3	3.3	89.2
Zhenjiang		17.7	37.6	28.5	1.1	3.1	2.6	1.8	87.4
Zhangjiagang		18.8	37.7	27.5	1.1	3.2	2.7	1.8	86.7

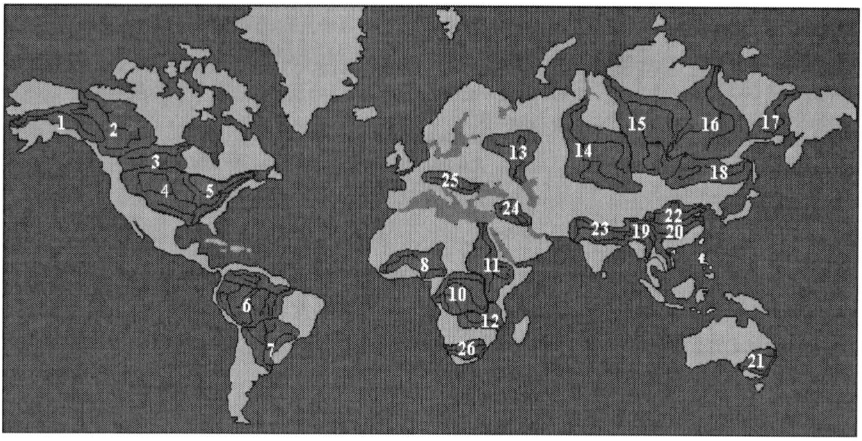

1. Yukon
2. Mackenzie
3. Nelson
4. Mississippi
5. St. Lawrence
6. Amazon
7. Parana
8. Niger
10. Congo
11. Nile
12. Zambezi
13. Volga
14. Ob
15. Yenisey
16. Lena
17. Kolyma
18. Amur
19. Ganges and Brahmaputra
20. Yangtze
21. Murryay Darling
22. Yellow River
23. Indus
24. Euphrates and Tigris
25. Danube
26. Orange

Figure 8.2. The drainage basines of the 25 majir world rivers.

CONCLUSION

For these thirty years after the arrest of the four gangs, Chinese scientists began to have the chance to do research works without unnecessary interruption.

Geochemists in China have been mainly engaged in obtaining fundamental and high quality information in relation to the distribution of nearly all the elements in the periodic table (Xie Ren and Yian, 2010). A large data base formerly named as Digital Element Earth was proposed (Xie et al., 2011a) and later changed as Chemical Earth (Wang, personal communications).

This database established will collect all data in the world and will be part of the activity of the on-going program of Global Scale Geochemistry. It will become a treasure of information for research works on theoretical geochemistry, exploration geochemistry, environmental geochemistry, petroleum geochemistry, and even for geology and metallogeny for many years to come.

REFERENCES

Bølviken, B., Stokke, P.R., Feder, J. & Jössang, T. 1992. The fractal nature of geochemical landscapes. *Journal of Geochemical Exploration*, 43 (2), 91–109.

Cameron, E.M., Hamilton, S.M., Leybourne, M.L., Hall, G.E.M., McClenaghan, B., 2004, Finding deeply-buried deposits using geochemistry. *Geochemistry: Exploration, Environment, Analysis* 4,7-32.

Cheng Hangxin, Shen Xiachu, Yan Guangsheng, Gu Tiexin, Lai Zhimin and Xie Xuejing, 1997. Wide-spaced floodplain sediment sampling covering the whole of China: pilot survey for international geochemical mapping. In: Xie Xuejing (Ed.), Geochemistry. *Proc. 30th Int. Geol. Congr. International Science Publishers*. 89-109.

Mann, A.W., 2010. Strong versus weak digestion: Ligand-based soil extraction geochemistry, *Geochemistry, Exploration, Environment, Analysis* V.10, 17-16.

Qian Desun and Xie Xuejing. 1953. Stream line operations of Large scale rapid chemical analysis. *Science Bulletin* (in Chinese), No.3:54-71.

Wang Xueqiu and Ye Rong 2011. Findings of Nanoscale Metal Particles: Evidence for Deep-penetrating Geochemistry (in Chinese), *Acta Geoscienica Sinica*, Vol 32 No.1: 7 12

Wensheng Yao, Xuejing Xie, Xueqiu Wang. 2011. Comparison of results analyzed by Chinese and European laboratories for FOREGS geochemical baselines mapping samples. GEOSCIENCE FRONTIERS 2(2) (2011) 247-259

Xie et al., 2008a Geochemical Atlas of 76 elements in SW China (in Chinese) 219PP, Geological Publishing House, Beijing.

Xie Xuejing (chief eaitor) 1986. Working Instructions, RGNR program (in Chinese) Geological Publishing House, Beijing. pp1-72

Xie Xuejing and Cheng Hangxin, 1997. The suitability of floodplain sediment as a global sampling medium. evidence from China. *Journal of Geochemical Exploration*, 58(1): 51-62.

Xie Xuejing and Ren Tianxiang, 1991. A Decade of Regional Geochemistry in China-the National Reconnaissance Project. Trans. IMM, 100: B57-B65.

Xie Xuejing and Sun Zhongjun, 2009 (in Chinese). A proposal of RGNR *Program for oil and gas*. Geological Bulletin of China.V.28, 1536-1538.

Xie Xuejing and Wang Xueqiu, 2003. New advances in deep-penetrating geochemistry (in Chinese). *Geosciences Fronties*. V.10 225-237.

Xie Xuejing and Yao Wensheng. 2010. Outlines of New Global Geochemical Mapping Program. *Acta Geologica Sinica*, 84 (3), pp.441-453.

Xie Xuejing and Yin Binchuan, 1993. Geochemical Patterns from Local to Global. *J. Geochem. Explor.*, 47: 109-129.

Xie Xuejing, 1977. Some problems in Regional Geochemical Mapping *Geophysical & Geochemical Exploration* (in Chinese) V.2, 1-10.

Xie Xuejing, 1979a. Regional Geochemical Mapping, 192 page, (in Chinese) Geological Publishing House, Beijing.

Xie Xuejing, 1979b. Discussion of methodology in RGNR program (in Chinese). *Geophysical and Geochemical Exploration* V.1, 18-26.

Xie Xuejing, 1995. New ideas and new technology in exploration of giant ore deposits. (in Chinese) scientific Chinese V.5.

Xie Xuejing, 2009, Surface geochemistry for oil and gas, what's its future (in Chinese). *Geological Bulletin of China*, V.28, 1533-1535.

Xie xuejing, Cheng Hangxin and Liu Dawen. 2008. Geochemical Mapping: With Special Emphasis on analytical requirements. *Acta Geologica Sinica*, 82 (2), pp.451-462

Xie Xuejing, Cheng Zhizhong, Cheng Hangxin 2004. Future prospect for the development of Applica *Geochemistry in China*. (in Chinese)

Xie Xuejing, Liu Dawen, Xiang Yunchuan, Yan Guangsheng and Lian Changyun, 2004. Geochemical blocks for predicting large ore deposit—concept and methodology. *J. Geochem. Explor*, 84:77-91.

Xie Xuejing, Lu Yinxiu, Yao Wensheng, Bai Jingfeng, 2011. Further study on deep penetrating geochemistry over the Spence porphyry copper deposit, Chile. *Geoscience Frotiners*, 2(3) : 303-311.

Xie Xuejing, Ren Tianxiang, Yan Guangsheng 2010. Road map for the Development of Exploration Geochemistry in the 21st century in China (in Chinese), *Geology in China*, V37. N.2, 245-267.

Xie Xuejing, Ren Tianxiang, Yan Guangsheng, Xi Xiaohuan, Liu Dawen, Wang Xueqiu, Cheng Zhizhong, Cheng Hangxin, Zhou Guanghua, Chi Qinghua, Sun Zhongjun 2010. Road-map of the, Development of Applied Geochemistry in China entering the 21st century. *Geology in China* (in Chinese). V.37, 245-267.

Xie Xuejing, Sun Huanzhen and Ren Tianxiang, 1989b. Regional Geochemistry-National Reconnaissance Project in China. *J. Geochemcal Exploration*, 33 (1-3): 1-9.

Xie Xuejing, Wang Xueqiu, Cheng Hangxin, Cheng Zhizhong and Yao Wensheng. 2011. Digital Element Earth. *Acta Geologica Sinica*, 85 (1), pp.1-16.

Xie Xuejing, Yan Mingcai, Li Lianzhong and Shen, 1985a. Geochemical Reference samples, Drainage sediment GSD1-8 from China. *Geostandard Newsletter*, 9(1):83-159.

Xie Xuejing, Yan Mingcai, Li Lianzhong and Shen, 1985b. Usable values from Chinese Standard reference samples of stream sediments, soils and rocks: GSD 9-12, GSS1-8 and GSR1-6. *Geostandard Newsletter*, 9(2):277-280.

Xie Xuejing, Yan Mingcai, Wang Chunshu, Li Lianzhong and Shen Huijun, 1989a. Geochemical Standard Reference Samples GSD9-12, GSS1-8 and GSR1-6. Geostandard Newsletter, V. XIII, No.1, 83-179.

Xie Xuejing. 2005. Foreword (for —The chemical compositions of the continental crust and rocks in the eastern part of China"). Science Press, Beijing, China.

Xie, Wang and Zhu, 2002 Comparison of multi-element analytical system and study of data quality monitoring scheme abstract in J.G.E.

In: Advances in Geochemistry Research
Editor: Jorge Sanjurjo-Sánchez
ISBN: 978-1-62618-245-5

Chapter 7

GEOCHEMISTRY OF CLAYS OF SURFICIAL MATERIALS FROM THE PALEOZOIC AND CENOZOIC OF REGUENGOS DE MONSARAZ REGION, OSSA MORENA ZONE

M. Isabel Dias*

ITN/IST, Instituto Superior Técnico,
Universidade Técnica de Lisboa, Sacavém, Portugal

ABSTRACT

Clay samples were collected in the framework of a minimization impact study of the Alqueva dam construction, in the Estremoz-Barrancos sector (SEB) of the Ossa Morena Zone, including the residual clays derived by weathering from diorites and gabbros, granodiorites and tonalites from the Reguengos de Monsaraz massif (RM), from schists, dolerite veins, vulcanites and schists with metabasites, and sedimentary clays (Tertiary), in the Reguengos de Monsaraz –Mourão – Póvoa region.

Mineralogical composition obtained by X-ray diffraction of both the bulk and <2 μm fraction made it possible to distinguish between some of those clay samples, but it was the chemical analyses of trace elements that played the most important role in obtaining regional geochemical

* E-mail address: isadias@itn.pt.

fingerprinting of both residual and sedimentary clays of surficial materials from the Paleozoic and Cenozoic of the studied area. REE patterns enhance the geochemical differences between clay samples, especially by the different degree of fractionation of REE and corresponding slope of REE pattern. The residual clay derived from the weathering of the dolerite vein was generally enriched in all elements, especially REE, and on the opposite end, the sedimentary Tertiary clays, due to the diluting effect made by the high amounts of carbonates, have the lower amounts of all chemical elements, especially REE, with the exception of As and Sb, whose ions are commonly fixed by clays and calcium. The clays from weathered diorites and gabbros, granodiorites and tonalites of RM have similar geochemical behavior as clays from weathered schists with metabasites, with high amounts of elements related with ferromagnesian minerals and HREE; LREE depletion is explained by the presence of heavy minerals like allanite. Clays derived by the weathering of schists have geochemical similarities with vulcanite clays, especially due to the high amounts of REE, and similar contents of Fe, Sc, Cr, even vulcanites are slightly enriched in Zn, As and Sb.

A good geochemical differentiation was achieved within and between residual clay materials and sedimentary clays of surficial materials from the Paleozoic and Cenozoic of the Reguengos de Monsaraz – Mourão - Póvoa region.

INTRODUCTION

The Portuguese part of the Ossa Morena zone (OMZ), a geotectonic unit of the Iberian Massif, was divided in three domains according to structural and stratigraphic criteria (Araujo et al, 2006), the Alter do Chão-Elvas sector, the Estremoz-Barrancos sector and the Montemor-Ficalho sector. The contact between these two sectors is done by faults. In the framework of studies to minimize impact of the Alqueva dam construction in the Guadiana basin (Portugal), a clay material survey was taken in the area, also including the area to be flooded (Dias et al., 2003; 2004; 2005a).

The main objective was to better identify and differentiate in a mineralogical and geochemical point of view, the variety of clay materials available in the Reguengos de Monsaraz – Mourão – Póvoa region. The data obtained from this work will contribute to other ongoing studies within the construction of the Alqueva dam, particularly for the characterization of raw materials and provenance of ancient pottery from archaeological sites in the impact area (Dias et al., 2012; 2010; 2007; 2005b; 2005c; Silva et al. 2010; Valera et al., 2007).

The studied area is included in the Estremoz-Barrancos sector (SEB), characterized by metamorphic and igneous rocks, comprising the sedimentary and volcano-sedimentary sequences of the Middle Cambrian to Lower Devonian, which is affected by late to post-tectonic Variscan deformations. The Reguengos de Monsaraz pluton is a late to post-tectonic Variscan intrusion with an irregular shape and a lithology mainly containing tonalites and granodiorites, and gabbro-dioritic rocks in the central part (Antunes et al, 2010).

In the south of this pluton, the Cenozoic Moura Basin area covers mostly of the Moura-Marmelar depression located north-east of Beja.

The metamorphic rocks of the OMZ comprise schists, greywackes, gneisses, amphibolites, metamorphised volcanic rocks, and metamorphic limestones, and the igneous rocks also have a variable lithology, including granites, granodiorites, quartz-diorites, tonalitic rocks, diorites, gabbros, charnokites, andesites (Araújo et al, 2006; Oliveira et al, 1991). It is important to enhance that geological interpretation of the area is sometimes difficult, due to a great complexity because of strong structural features. Low, gently hilly lands are the dominant landscape element, with altitudes ranging from 200 to 250 meters.

This polygenic erosion surface cuts through the different lithologies (metamorphic and igneous rocks) with Tertiary deposits standing over some of these formations.

The Tertiary deposits from Moura-Marmelar basin (Moura) in the eastern sector, in Amareleja and in Safara sub-basins, are represented by the Upper Miocene depositional sequence which is related to a fluvial depositional system - a pre- Guadiana river draining to the west (Pais et al, 2012; Cunha et al, 2009); these fluvial deposits include siliciclastic sandstones and siltites with some interbedded conglomerates (quartz and schist pebbles).

An important feature occurs in the eastern area, culminating this fluvial depositional system, represented by palustrine carbonate deposits and related to endorheic drainage, and often also pedogenic carbonate accumulations (calcretes) (Brum da Silveira, 1990; 2002).

Clay samples were collected in this area comprising the weathered materials of the lithologies from SEB metamorphic and igneous rocks, as well as, the sedimentary Tertiary clays.

A chemical characterization by neutron activation analysis was performed on the bulk material and also a mineralogical characterization by X-ray diffraction on both bulk and <2 μm fraction.

METHODS AND MATERIALS

Chemical analyses were done by instrumental neutron activation analyses (INAA), determining major and trace elements contents with very good accuracy and precision (in general <5%). Bulk clay samples were ground into a fine powder in an agate mortar and then dried in an oven at 100 °C for 24 hours and stored in desiccators until the samples can be weighed. The following elements were determined: Na, K, Fe, Sc, Cr, Co, Zn, As, Ga, Br, Sb, Rb, Cs, Ba, La, Ce, Nd, Sm, Eu, Tb, Yb, Lu, Hf, Ta, Th, U. Samples and standards were irradiated together in the core grid of the Portuguese Research Reactor (ITN, Sacavém) for seven hours (longer irradiation), at a thermal neutron flux of 4.0E12 n/cm2/s, epithermal 3.7E10 n/cm/s and fast flux of 1.6E10 n/cm2/s. The gamma-ray analysis was performed using a Ge γ spectrometer consisting of a 150 cm3 coaxial detector and a low energy photon detector (LEPD), connected through Canberra 2020 amplifiers to Accuspec B (Canberra) multichannel analyser. This system has a FWHM of 1.9 keV at 1.33 MeV (coaxial Ge detector), of 300 eV at 5.9 keV and of 550 eV at 122 keV (LEPD). Corrections for the spectral interference from U fission products in the determination of Ba, REE and Zr were done according to Gouveia et al. (1987) and Martinho et al. (1991). More details of the analytical method were published elsewhere (Prudêncio et al, 1986; Dias and Prudêncio, 2007).

A multivariate statistical analysis of chemical data was done by using the Statistica data analysis software system (StatSoft, Inc., 2011).

The mineralogical analysis was obtained by X-ray diffraction (XRD), using a Philips diffractometer, Pro-Analytical, with Kα Cu radiation, for powders of non-oriented aggregates on the bulk sample, scanned from 2 to 70° 2θ and a step size of 1° 2θ/min. Oriented specimens of the <2 mm fraction were prepared by air drying a small amount of dispersed suspension onto a glass slide.

To help identify the clay minerals, in addition to the sample preparations described above, ethylene glycol-solvated (8 h) samples and heated samples at 550°C for 2 h were also prepared using well known methods (Thorez, 1976; Brindley and Brown, 1980; Moore and Reynolds, 1997).

Clay samples collected include the residual clays derived from the weathered schists, diorites and gabbros, granodiorites and tonalites, dolerite veins, vulcanites and schists with metabasites, as well as to the sedimentary clays of the Tertiary deposits, in the region of Reguengos de Monsaraz – Mourão - Póvoa (Figure 1).

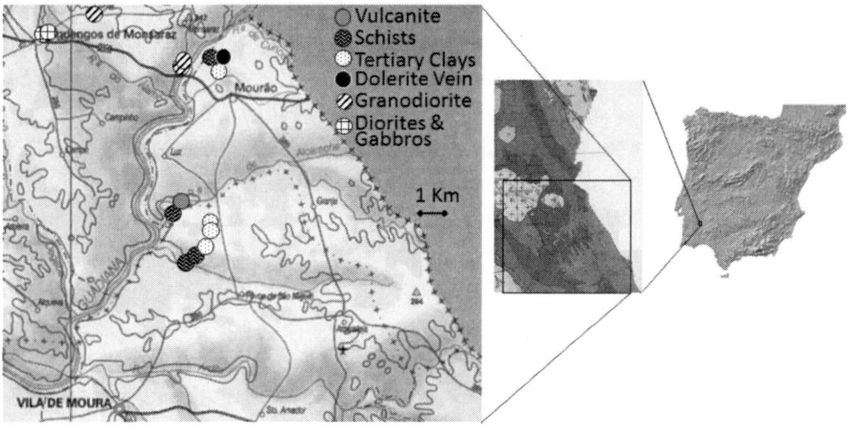

Figure 1. Clay samples location.

RESULTS AND DISCUSSION

Mineralogical Composition

Clay materials in the studied sector of OMZ are mainly residual clays resulting from the weathering of the various lithologies existent in the area, and also sedimentary clays belonging to the Cenozoic cover of the region.

All the lithologies of the Reguengos de Monsaraz massif show textural and mineralogical heterogeneities marked by the widespread occurrence of mafic microgranular enclaves throughout the dominant granitoid types and the presence of irregular patches of quartz-dioritic and tonalitic composition within the gabbro-diorites (Antunes et al, 2010).

Residual clay materials of gabbro-diorites and granodiorites have in general plagioclase as the major rock forming mineral. From the gabbro-diorites to the tonalites and the granodiorites we assist to an increase of the albite content in the plagioclases, together with an increase of microcline and quartz proportions and a decrease of mafic mineral contents (specially, amphibole); quartz is always present. Amphiboles are the main ferromagnesian minerals, and as accessory minerals iron oxides were also identified by XRD. Depending on the weathering degree of the sampled clay materials, the phyllosilicates content differs.

Vulcanite weathered sample has quartz as dominant mineral, associated to phyllosilicates and plagioclase, with traces of k-feldspar and iron oxides.

The schists weathered products have as bulk mineralogical association quartz + phylosilicates (+calcite) + traces of feldspars and iron oxides. It is important to enhance that due to the geological complexity of the region, calcoschist rocks are also present, as well as, recrystalized carbonates affected by schistosity, and so some –schistous" weathered samples are calcite rich.

The clay sample collected in a weathered dolerite vein which cuts the schists has higher amounts of phyllosilicates, followed by quartz, and in lower amounts calcite, k-feldspars, plagioclase and hematite.

The Cenozoic Moura Basin area covers mostly the Moura-Marmelar depression located north-east of Beja. Several isolated and smaller sub-basins are also included in the Moura Basin, interpreted on the base of equivalent stratigraphy, similar sedimentary characteristics and the same morphotectonic context (Pais et al, 2012). The existing lithostratigraphic model controls the mineralogical composition which reflects the several environmental contexts and related depositional sequences. Four unconformity bounded depositional sequences were identified (Pais et al, 2012) and are mainly related to alluvial fan systems (conglomerates, sandstones), palustrine systems (limestones), fluvial systems (sandstones, conglomerates, claystones), and two major episodes of pedogenic carbonate development (calcretes) (Brum da Silveira, 2002). Bulk sample mineralogy of Tertiary analysed clays comprises calcite associated to quartz and phyllosilicates, with traces of feldspars. Nevertheless the calcite amounts may vary from dominant (close to Mourão and Monte dos Trincalhos) to trace amounts (close to Maganos and Altas Moras, north of Póvoa) reflecting the heterogeneity of this Tertiary deposit and the influence of the episodes of the development of carbonates in some units.

Regarding the clay fraction mineralogy, the residual clays from the weathered granodiorites and tonalites, the schists with metabasites and the dolerite vein, as well as the sedimentary Tertiary clays samples, have the same clay minerals association, with higher amounts of smectite associated to illite and traces of kaolin minerals. Nevertheless, Tertiary clays can be differentiated by the presence of ordered smectite (especially enhanced in the more calcite rich ones). The diorites and gabbros clay materials have a clay fraction rich in illite with some smectite (degraded clay minerals) and in a few cases probably also traces of pyrophilite. The residual clays derived from the weathered schists have clay fractions also rich in illite, but associated to chlorite, and traces of kaolin minerals and of smectite. The clay fraction from the weathered vulcanite is chiefly illite followed by kaolin minerals and smectite (very disordered) and also interstrafied illite/smectite, illite/chlorite and traces of chlorite. It is important to enhance the difficulty in differentiating residual clays based on the mineralogical composition.

Chemical Composition

Chemical composition of clays derived from the weathered sampled materials is evidently correlated with parent-rocks. As mentioned above in the methods and materials section, the chemical results presented and discussed here correspond to the bulk sample.

The clays derived from the weathered granodiorites and associated tonalites, diorite and associated gabbros have a mafic behaviour (Cluster 3, figure 2) generally with higher amounts of Fe, Sc, Cr, Co (Figure 2). They also have higher amounts of heavy rare earth elements (HREE) relatively to the light rare earth elements (LREE). Some scattered variation was found, especially in some trace elements (e.g., Zr, Hf, REE) pointing to variations within the Reguengos de Monsaraz (RM) massif where they are included, what may be explained by the fact that the gabbros and diorites are the more mafic terms in contrast with the more felsic terms of the tonalites and granodiorites. Indeed the behaviour of such elements is strongly controlled by accessory minerals like zircon (Zr, Hf) and allanite (REE), which presence also justifies the fractionation of the HREE relatively to the LREE.

The clays derived from the weathered schists have also slightly high amounts of K, Fe, Sc, Cr, Co, Zn and the higher Rb, Zr, Hf, Ta, Th, U and REE contents (Cluster 2, Figure 2). The enrichment in K, Rb, Ga may be explained by the importance of phyllosilicates in parent rocks and by its terrigenous origin. Previous works (Antunes et al, 2010) already enhance the enrichment of REE in these schists, mainly explained by the lithogeochemistry of the Ordovician, as the quartzites interstratified in the schists, immediately above the thick base quartzites, have the higher amounts of rare earths, and not the schists itself.

The clay derived from the weathered vulcanite has high amounts of Zn, As, Sb and HREE, which are generally dispersed in clays, hydrous Fe and Mn oxides, pyroxene, and other sulphides and phosphates.

The residual clay of the weathered dolerite vein has in general high contents of all elements, with the exception of Na, and also Fe, Sc, Zn and Rb, that are in low amounts in the sample, but don‘t present the lowest amounts of all analyzed samples (Figure 3). It is important to enhance that this sample is detachable from the others especially due to the higher contents of K, Cr, As,

Zr, Sb, Cs, REE (especially LREE) and Hf. These elements are usually concentrated in clays, iron oxides and heavy minerals, being in accordance with the XRD results of the studied dolerite vein clay sample.

Tertiary clay samples (Cluster 1, Figure 2) stand out due to higher carbonate contents, higher As and Sb, and lower Fe, Sc, Cr, Co, and REE contents (but LREE>HREE). The high calcium carbonate amounts justifies the higher Ca content, and also the higher As content of this samples, as arsenate ions are readily fixed by clays and calcium.

REE patterns enhance the geochemical differences between all clay samples analyzed, especially by the different degree of fractionation of REE and corresponding slope of REE pattern (Figure 4). The clays derived from the weathered granodiorites and tonalites and from the diorite and gabbros of the RM massif have in general moderate enrichment in LREE ($5.17 \leq LaN/YbN \leq 6.40$) and slightly higher enrichment in HREE; this somewhat LREE depletion is certainly related to the presence of heavy minerals like allanite in these samples (as previously reported by petrographic studies of Antunes et al 2010).

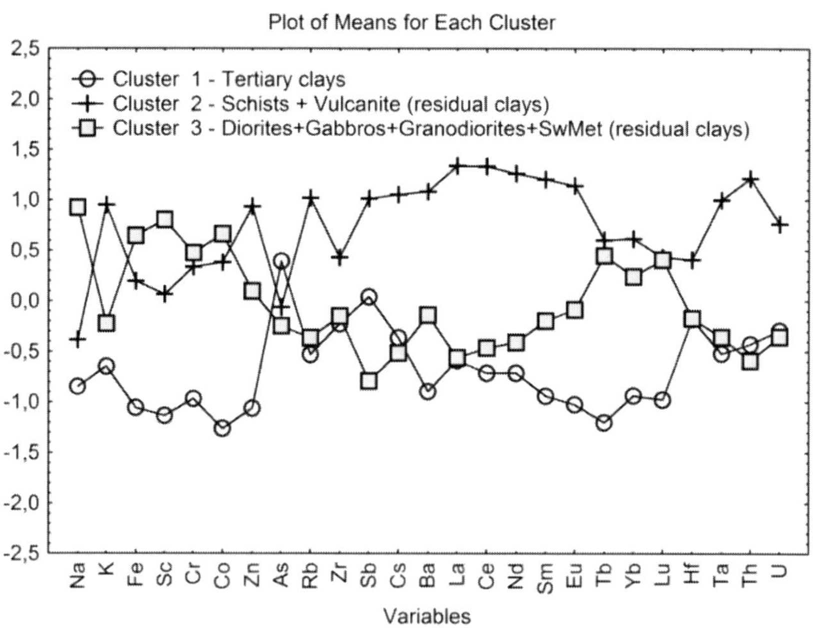

Figure 2. Plot of means for each cluster obtained by the k-mean clustering method, using as variables the major-, minor-, and trace-element concentrations obtained by INAA on the bulk material of clay samples, excluding the dolerite vein residual clay.

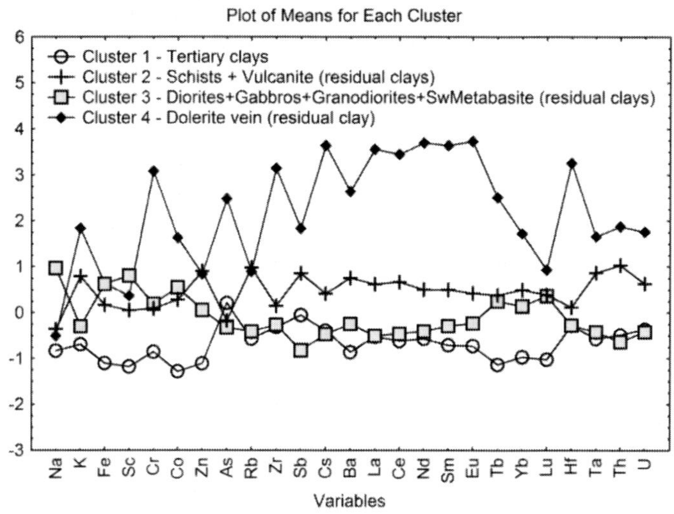

Figure 3. Plot of means for each cluster obtained by the k-mean clustering method, using as variables the major-, minor-, and trace-element concentrations obtained by INAA on the bulk material of all clay samples. This figure intends to emphasise the chemical behaviour of the clay sample derived from the weathered dolerite vein.

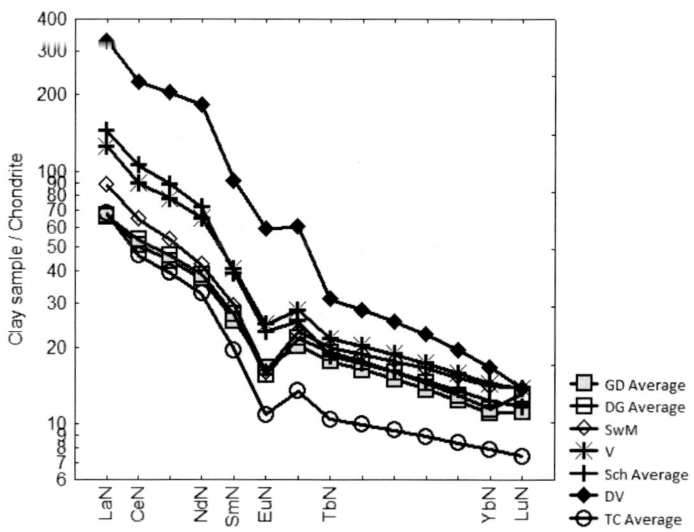

Figure 4. REE distribution patterns obtained for the average values of clay materials relative to the chondrites (Haskin *et al.*, 1971). Residual clays: GD- Granodiorites; DG-GabbrosandDiorites; SwM-Schist with metabasite; V-vulcanite; Sch-Schist; DV-dolerite vein; Sedimentary clays: TC-Tertiary clays.

They also have negative Eu anomalies of variable magnitude ($0.65 \leq Eu/Eu^* \leq 0.81$) and have similar sum REE (granodiorites and tonalites = 98.23; diorite and gabbros = 102.91).

The clay materials from the weathered schists with metabasites have a similar REE pattern, with moderate enrichment in LREE ($La_N/Yb_N = 6.25$), slightly high enrichment in HREE and the presence of negative Eu anomaly ($Eu/Eu = 0.61$) and a sum REE=121.95. On the other hand the clays derived from the weathered schists have higher contents of REE (sum REE = 195.72), especially in LREE ($10.30 \leq La_N/Yb_N \leq 21.73$), and an extreme depletion of the HREE relative to the LREE, probably due to the presence of garnet in the source. The negative Eu anomalies have also various magnitudes ($0.62 \leq Eu/Eu^* \leq 0.85$). The residual clays from the dolerite vein are the most enriched in REE (Sum REE = 441.57), especially the LREE (La_N/Yb_N=19.66) and have an Eu anomaly of 0.83. The Tertiary clays have the lower REE concentrations of all analyzed clay samples (sum REE = 89.33), what might be explained by the fact that they are carbonate rich, which cause a diluting effect on REE. They have a moderate enrichment in LREE ($7.23 \leq La_N/Yb_N \leq 10.48$) and negative Eu anomalies ($0.58 \leq Eu/Eu^* \leq 0.69$).

CONCLUSION

Field work was done in the region Reguengos de Monsaraz – Mourão – Póvoa included in the Estremoz-Barrancos sector (SEB) of the Ossa Morena Zone (OMZ), in the framework of the interdisciplinary studies to minimize impact of the Alqueva dam construction in the Guadiana basin (Portugal). In this work the main purpose was to establish a good mineralogical and geochemical differentiation of clays of surficial materials from the Paleozoic and Cenozoic of Reguengos de Monsaraz region, OMZ. The more clayey materials of the region are the sedimentary clays from the Tertiary deposits and the residual clays derived from the weathered (i) diorites and gabbros and (ii) granodiorites and tonalites from the Reguengos de Monsaraz massif (RM); (iii) schists; (iv) dolerite veins (v) vulcanites and (vi) schists with metabasites.

The foremost obtained results of the mineralogical composition of both the bulk and the <2μm fraction of residual and sedimentary clays from the Reguengos de Monsaraz – Mourão – Póvoa region may be summarized as follow:

1) differentiation of Reguengos massif (RM) clay materials, particularly between more siliciclastic and smectite (disordered) rich materials (granodiorites and tonalities clays and schists with metabasites clays) from siliciclastic, amphibole and illite (+smectite) clay materials (diorites and gabbros);
2) the vulcanite clay sample may be described as ―common clay”, siliciclastic, very clayey with illite, smectite (very disordered), kaolin minerals, interstratified illite/smectite, illite/chlorite and traces of chlorite;
3) clay materials derived from weathered schists have a general mineralogical association of quartz + phylosilicates (+calcite) + traces of feldspars and iron oxides; clay minerals association mostly contains illite, chlorite and traces of kaolin minerals and of smectite; in some cases the clay materials derived from weathered schists are calcite rich, most probably due to the weathered products of calcoschist rocks or recrystalized carbonates affected by schistosity, that are also present in the geological complexity of the region;
4) Tertiary clays are calcite rich and the clay minerals are smectite (ordered) associated to illite (and traces of kaolin minerals).

In a geochemical point of view, and complementing mineralogical obtained data, an improved differentiation was achieved within and between residual clay materials and sedimentary clays of Reguengos de Monsaraz – Mourão - Póvoa region:

1) the clays derived by weathering of the lithologies of the Reguengos de Monsaraz massif (granodiorites and tonalities, and gabbro-diorites) have an intense LREE depletion due to the presence of heavy minerals like allanite, high contents of elements related to ferromagnesian minerals, and an increase of albite in plagioclase and of microcline and quartz, and a decrease of mafic minerals like amphibolite, from the gabbro-diorite term to the granodiorites and tonalities term;
2) the clays derived by the weathered schists have high amounts of K, Fe, Sc, Cr, Co, Rb, REE, Ta, Th and U; the clearly enrichment in K, Rb, Ga may be explained as a consequence of the importance of phyllosilicates in parent rocks and of its terrigenous origin; high contents of REE may be explained by their richness in Ordovician parent rock and interstratified quartzites in these schists, and not the schists itself;

3) the clay rich material resulting from the weathered dolerite vein has in general high contents of all elements, related to their preference to be fixed in clays, iron oxides and heavy minerals;
4) Tertiary clays have a general depletion of all elements, especially REE, as a diluting effect of the presence of high amounts of carbonates in these samples, and are enriched especially in As, as arsenate ions tend to be readily fixed by calcium.

The identification of prevailing clay materials in the studied region (Reguengos de Monsaraz – Mourão – Póvoa) was achieved, and a good mineralogical, and especially geochemical, differentiation was attained, in the variety of both residual and sedimentary clay materials. In this way, geochemical and mineralogical characterization of clays of surficial materials from the Paleozoic and Cenozoic of Reguengos de Monsaraz region (OMZ) is now available, and is particularly important to complement other studies, especially those performed in the framework of Alqueva dam minimization impact works.

ACKNOWLEDGMENTS

The financial support of the Portuguese Foundation for Science and Technology (FCT/MCTES) is acknowledged through the Project PTDC/CS-ANT/104333/2008.

The assistance in the fieldwork campaign is acknowledged to Era Arqueologia SA.

REFERENCES

Araújo, A.; Piçarra Almeida, J.; Borrego, J.; Pedro, J. and Oliveira, T. (2006). As regiões central e sul da Zona de Ossa Morena. In: Geologia de Portugal no Contexto da Ibéria (Dias, R.; Araújo, A.; Terrinha, P. and Kullberg, J.C., eds.), Universidade de Évora, 151-172.

Brindley, G.W. and Brown, G. (1980) Crystal Structures of Clay Minerals and their X-ray Identification. Monograph 5, Mineralogical Society, London.

Brum da Silveira A (1990) Neotectónica e Sismotectónica da Região Vidigueira–Moura. MSc, Univ. Lisboa, p 204.

Brum da Silveira A (2002) Neotectónica e sismotectónica de um sector do Alentejo oriental. Ph.D., Univ. Lisboa, p 339.

Cunha PP, Pais J, Legoinha P (2009) Evolução geológica de Portugal continental durante o Cenozóico-sedimentação aluvial e marinha numa margem continental passiva (Ibéria ocidental). 68 Simposio sobre el Margen Ibérico Atlántico MIA09, pp 11–20.

Dias, M.I. and Prudêncio, M. I. (2007) "Neutron activation analysis of archaeological materials: an overview of the ITN NAA Laboratory, Portugal. *Archaeometry*, vol 49, 2, p.381-391.

Dias, M.I., Prudêncio, M.I., Trindade, M.J., Valera, A.C. (2012). ―Towards a Temporality Approach in Perdigões, Portugal: Chemical and Mineralogical Composition of Neolithic and Chalcolithic Pottery and Raw Materials". Macla nº 16. junio '12, Sociedad Española de Mineralogía, pp. 28-29.

Dias, M.I., Valera, A.C., Lago, M. l., Prudêncio, M.I. (2007): Proveniência e tecnologia de produção de cerâmicas nos Perdigões. *Vipasca Arqueologia E História.* 2. 2ª sér. 2007. 117-121

Dias, M.I., M. I. Prudêncio, F. Rocha, (2005a) – ―Geoquímica e mineralogia de matérias-primas não metálicas: aplicação a estudos de cerâmicas arqueológicas na área da barragem do Alqueva, Portugal." Actas do XIV Semana de Geoquímica e VIII Congresso de Geoquímica dos Países de Língua Portuguesa. Pp. 289-292.

Dias, M.I., Valera, A.C., Lago, M., Gouveia, M.A. (2005b): Composition, Technology and functional features of Chalcolhitic pottery from Perdigões, Reguengos de Monsaraz (Portugal). Geoarchaeological and Bioarchaeological Studies, 3, 161-164.

Dias, M.I., Albergaria, J., Jorge, A.A., Ramos, C., Melro, S., Prudêncio, M.I. and Rocha, F. (2005c). ―Ceramic production in Monte da Pata and Castelo das Juntas Late Iron Age archaeological sites (Guadiana Basin, Southern Portugal): some archaeological results". In: *Understanding People Through their Pottery.* EMAC'03. Trabalhos de Arqueologia, Série Monográfica Trabalhos de Arqueologia, nº 42, IPA, M.I. Dias, M.I. Prudêncio and J.C. Waerenborgh, eds., pp. 51-60.

Dias, M.I., Rocha, F., Prudêncio, M.I.; Valera, A. C. (2004). Mineralogy and geochemistry of raw materials applied to ceramic studies from Pre-history to Roman times, in the Alqueva dam impact area, Portugal, 32nd International Geological Congress, Florence, Italy, August 20-28, 2004. Book of Abstracts, part 1., p. 368.

Dias, M.I., Rocha, F., Prudêncio, M.I. (2003). Clay materials exploitation in the Alqueva dam region (Portugal) in ancient times, Euroclay, Modena, Italy, 2003. *Book of Abstracts*.

Gouveia, M.A., Prudêncio, M.I., Freitas, M.C., Martinho, E., Cabral, J.M.P. (1987). Interference from uranium fission products in the determination of rare earths, zirconium and ruthenium by instrumental neutron activation analysis in rocks and minerals." *Journal of Radioanalytical and Nuclear Chemistry,* 2 (Articles 14), 309-318.

Haskin, L.A., Helmmke, P.A., Paster, T.P. and Allen, R.O. (1971). Activation Analysis in Geochemistry and Cosmogeochemistry (A.O.Brunfelt and E.Steinnes, editors), Universitetsforlaget, Oslo, 201-218.

Martinho, M. A., M. A. Gouveia, M.I. Prudêncio, M.F. Reis, J.M.P. Cabral (1991). Factor for correcting the ruthenium interference in instrumental neutron activation analysis of barium in uraniferous samples. *Applied Radiations and Isotopes*, 42, 1067-1071.

Moore, D. M., Reynolds, R.C. (1997). *X-Ray Diffraction and the Identification and Analysis of Clay Minerals.* Oxford University Press. 2nd edition. 378p.

Oliveira, J.T.; Oliveira, V. and Piçarra, J.M. (1991). Traços gerais da evolução tectono-estratigráfica da Zona de Ossa Morena em Portugal: síntese crítica do estado actual dos conhecimentos. Comunicações dos Serviços *Geológicos de Portugal*, 77: 3-26.

Pais, J., Cunha, P., Pereira, D., Legoinha, P., Dias, R., Moura, D., Brum da Silveira, A., Kullberg, J., Gonzalez-Delgado, A. et al., (2012) *The Paleogene and Neogene of Western Iberia* (Portugal). A Cenozoic record in the Erupoean Domain. SpringerBriefs in Earth Sciences. 155 p.

Prudêncio, M.I., Gouveia, M.A., and Cabral, J.M.P. (1986) Instrumental neutron activation analysis of two French geochemical reference samples basalt BR and biotite Mica-Fe. Geostandards Newsletter, X, 2931.

Silva, Ana Maria, Valera, António Carlos, Costa, Cláudia E Dias, Maria Isabel (2010), A new research project on funerary practices at Perdigões enclosure. Apontamentos de Arqueologia e Património – 5 / 2010, pp. 43-48.

Thorez, J. (1976) *Practical Identification of Clay Minerals*. G. Lelotte, Dison, Belgium.

Valera, A., Lago, M.; Duarte, C., Dias, M.I., Prudêncio, M.I. (2007): "Investigação no complexo dos Perdigões: ponto da situação de dados e problemas", Actas do 4.° Cong. Arqueologia Peninsular, Braga, Universidade do Algarve. Universidade do Algarve. Promontoria Monográfica 08. Pp. 53-66.

INDEX

A

B

C

D

G

H

I

J

L

M

N

O

P

Q

R

S

T

U

V

W

X

Z